CASE STUDIES IN ENVIRONMENTAL TECHNOLOGY

CASE STUDIES IN ENVIRONMENTAL TECHNOLOGY

Edited by Paul Sharratt and Michael Sparshott

INSTITUTION OF CHEMICAL ENGINEERS

Published by
Institution of Chemical Engineers,
Davis Building,
165–189 Railway Terrace,
Rugby, Warwickshire CV21 3HQ, UK.

© 1996 Institution of Chemical Engineers
A Registered Charity

ISBN 0 85295 385 2

The symposium upon which this book is based was organized by the Institution of
Chemical Engineers North Western Branch and the Environmental Protection Subject
Group. The event was co-sponsored by AEA Technology, the Institution of Civil
Engineers North Western Association, the Institution of Mechanical Engineers
Environmental Engineering Group, the Royal Society of Chemistry Environmental
Chemistry Group and the Society of Chemical Industry Environment and Water Group.

THE INSTITUTION OF
CIVIL ENGINEERS

AEA Technology

I MECH E
150th Anniversary
1847-1997

Cover photograph: microwave-energized plasma for VOC destruction (courtesy of EA
Technology Ltd)
Printed in the United Kingdom by Redwood Books, Trowbridge, Wiltshire

ii

PREFACE

This book describes a range of emerging and recently established techniques and technologies for process improvement, waste treatment and the monitoring of environmental performance.

The impact of a process on the environment can be reduced by changes within the process itself, by better or more efficient techniques to minimize and render harmless emissions to air, land and water, or by a combination of both. Where improvements within the process can be achieved, process cost reductions often also result, while the associated emissions reduction may reduce the cost of waste disposal, be this on-site (end-of-pipe) treatment or off-site (contractor) disposal. The first five chapters look at ways of identifying such process improvements during design and operation of processes.

Chapter 1 addresses all aspects of emissions reduction at a large, complex chemicals site, emphasizing water management issues. Detailed surveys of water use led to lower water consumption, aqueous effluents reduction, better treatment performance and reduced utility costs. Chapter 2 considers themes of the '3Es' — emissions, efficiencies and economics — and the benefits of a systematic examination of all processes on a site. Attention to detail in completing mass balances is another way of pinpointing losses and potential savings. Chapter 3 looks at a formal system for solvent management that can be used by solvent users of any size. The benefits of applying lateral thinking to plant and process reliability are demonstrated in Chapter 4. Identification of sources of spillage problems generated costs savings — not only in the costs of providing secondary containment to satisfy the regulators (and with their agreement), but also in the costs of operation. Chapter 5 identifies some of the problems specific to batch processes, providing a range of strategies that may reduce both waste and costs.

The next two chapters address two aspects of monitoring. Irrespective of the means of waste disposal, emissions must be measured to allow quantitative regulation. In the UK, Integrated Pollution Control (IPC) has produced a plethora of monitoring requirements, with a preference for continuous monitoring. Chapter 6 looks at the UK regulators' approach to standardization of monitoring techniques, based on the national and international standards that exist for measurement methods. This approach will cause problems to multinational companies with the need for comparable performance testing worldwide, and who have traditionally developed methods internally. It is clearly necessary as a step towards a level playing field in environmental performance monitoring, however, as well as in the development of instrumentation. Chapter 7 looks at an emerging monitoring technique that may have applications in the difficult area of odour pollution. By use of modern sensor technology, instruments have been developed that could provide a fast, quantitative and reliable alternative to the infamous 'odour panel'.

Traditionally, the approach to design of effluent treatment systems has been to mix wastes together and use a single treatment plant. Chapter 8 challenges this view, and shows how a systematic application of the pinch method can identify opportunities for money-saving distributed effluent treatment.

New technologies also have a part to play, both in reducing waste at source and in abating emissions to the levels required by the regulators. The remaining chapters address various technologies for the avoidance or treatment of wastes and for resource recovery. Chapter 9 shows how NO_x emissions from discrete and specialized processes such as nitration and metals pickling can be reduced or eliminated, either by destroying the NO_x as part of the process, or by eliminating the use of nitric acid — surely an ideal solution. The use of new materials extends the capabilities of membrane separations to deal with extremes of pH and systems with aggressive solvents. Chapter 10 shows how this can be exploited to recycle troublesome wastes such as caustic washes.

A serious bone of contention recently has been the use of combustible waste as fuel by the cement industry. This issue is explored from various perspectives in Chapters 11 to 13. Traditional waste incineration industries have suffered, although they will be essential in dealing with many difficult wastes, and the regulators have had a troubled time over authorizations. The cement industry, on the other hand, is pleased with the improved performance that it can demonstrate against the Environment Agency's Best Practicable Environmental Option (BPEO) guidance. Paradoxically, the solvent recovery industry may benefit from this solvent waste disposal route — the availability of a cost-effective sink for still residues means that more difficult recoveries can be carried out profitably.

Capture of volatile organic compounds (VOCs) by cryogenic condensation is a relatively new technology described in Chapter 14. It is particularly attractive when the cooling is provided at no additional cost from a site's liquefied nitrogen supplies. Techniques for the destruction of VOCs in gas streams now include a microwave-induced plasma. Chapter 15 describes how atmospheric pressure plasmas can be created using equipment similar to a domestic microwave, giving an impressive VOC destruction capability.

Vapour-phase treatment using flameless thermal oxidation is described in Chapter 16, which presents data on the destruction of various recalcitrant chlorinated components of solvent wastes from a range of industries. Simple products such as hydrogen chloride can be removed by scrubbing. Finally, Chapter 17 discusses the effective and well-established (yet still developing) applications of activated carbon for the clean-up of vapour streams.

We would like to thank others who have been involved in the preparation of this book. In particular, Ian McConvey and Tony Thompson are owed much for their energy and enthusiasm in both organization and soliciting contributions.

Paul Sharratt
Mike Sparshott

THE AUTHORS

CONTENTS

1. WASTE MINIMIZATION AT A LARGE INTEGRATED CHEMICAL MANUFACTURING SITE

Simon Clouston and Peter Shields

The 'true' cost of waste is often several times larger than the cost of waste disposal. The 'true' cost includes the value of raw materials or products lost in waste streams as well as the over-consumption of utilities (for example, water, steam, nitrogen) and consumables (for example, water treatment chemicals). For the Associated Octel site at Ellesmere Port the 'true' cost of waste is more than £10 million per year. An initial review of waste arisings on the site identified over 150 different types of waste.

The waste minimization approach has initially been applied to water. The site currently uses more than 5,000,000 m^3 of water per year, approximately 60% of which is discharged to the Manchester Ship Canal.

This chapter describes how a systematic approach to waste minimization is being implemented at Associated Octel.

INTRODUCTION

The Associated Octel site at Ellesmere Port has manufactured motor fuel anti-knock compounds (MFAKC) since the 1950s. The Ellesmere Port site also produces many of the chemicals that are subsequently used in the manufacture of MFAKC, including sodium (subsequently used on-site to manufacture lead-sodium alloy), and chlorine and hydrogen (subsequently used to manufacture ethyl chloride).

The world market for MFAKC products is in decline and is unlikely to exist in any significant form in 15–20 years' time. For this reason the Ellesmere Port site is increasingly moving towards the manufacture of organic fine chemicals. The detailed attention to operating costs that comes with manufacturing in a declining market and the requirements of the Environment Agency (the Ellesmere Port site has eight IPC authorizations) convinced Associated Octel that the time was right to adopt a systematic approach to waste minimization.

WS Atkins has worked extensively with Associated Octel over a number of years on a variety of projects, including preparation of IPC applications, front end design studies and assessments of existing processes. WS Atkins also has considerable waste minimization expertise, having undertaken numerous projects for industrial clients as well as managing regional demonstration projects such as:

- Project Catalyst — 14 companies in North West England;
- Dee Project — 14 companies in North East Wales;
- IRTU (WEFT Project) — 8 companies in Northern Ireland;
- Medway and Swale Project — 16 companies in Kent.

As a result of this background WS Atkins was asked to assist Associated Octel with the implementation of a waste minimization programme at the Ellesmere Port site.

Aerial view of the Associated Octel site at Ellesmere Port.

WASTE ARISINGS REVIEW

The first stage was to carry out a review of the waste arisings across the site. The purpose of this review was to produce a short-list of waste from which Associated Octel could select one waste issue to be the subject of a pilot waste minimization project for the site. The main criterion for inclusion on the short-list was the significance of the 'true' cost of that waste, or more precisely the estimated potential savings that would accrue.

The list of waste arisings was generated at a series of brainstorming sessions with key production, maintenance, engineering and technical support staff from all manufacturing areas as well as the work services and engineering groups.

These sessions were followed by further discussions with key personnel from these areas to obtain data on the overall annual costs and quantities for the wastes identified. Wastes were identified as inputs to the business — that is, purchased goods and services — rather than effluent discharges. This enabled a much better estimate of the 'true' cost of waste to be developed. The data gathered at this stage did not need to be very accurate; just sufficient to enable the various wastes to be ranked in relation to one another. Also, time was not wasted pursuing cost and quantity data on wastes that were clearly not significant overall. These waste items included gaskets, canteen waste, scrap welding rods and so on. Table 1.1 gives examples of the range of wastes identified, with typical cost and quantity data. The heading 'resource type' is used because much of the data collected at this stage relates to total consumption and obviously not all of this is waste.

TABLE 1.1
Typical data obtained during the review of one plant area

Resource type	Quantity, tonnes per year	Value, £ per year
Water	630,000	176,000
Solvent	35	12,250
Graphite	300	90,000
Disposable PPE	–	350,000
Spent absorber oil	15	6000
Spent molecular sieve	1	–

TABLE 1.2
Short-list of wastes presented to site Group Managers

Waste type	Target savings, £ per year
Electricity	800,000
Steam	500,000
Water	800,000
Caustic (absorption plant make-up)	120,000

In total over 150 different types of waste were identified and cost and quantity data collated for more than 80 of these. The top ten wastes, in terms of cost, were listed; they included raw materials and product losses as well as utilities.

For some of these wastes projects were already in progress, so these were excluded from further consideration. Table 1.2 shows the final short-list. This was presented to the Manufacturing and Engineering Group Managers with the recommendation, approved after debate, to select site water usage as the subject of the pilot waste minimization project. The reasons for selecting water were:

- the cost of water was significant (£1,600,000 in 1995);
- evidence from many other sites demonstrates that significant reductions can be achieved;
- all site personnel can relate to the over consumption of water and the generation of aqueous effluent as waste;
- there are usually some improvements that can be identified and implemented rapidly so success can be demonstrated early;
- it would aid compliance with a target in one of the site's IPC authorizations — that is, to reduce aqueous discharges to the Manchester Ship Canal by 50%;
- the use of water is at the heart of many processes and studying it can lead to benefits in other areas.

SCOPE OF THE WATER PROJECT

The agreed scope of the pilot project was all water use on the Ellesmere Port site. The overall target was for a reduction in water consumption of 50%. Most of the water used on site (90–95%) is abstracted from the River Dee. The balance is mainly supplied by towns water with a small contribution from water abstracted under licence from the Manchester Ship Canal.

Following approval by the Group Managers to focus on water usage, the approach to this next stage of the project was developed to:

- make most efficient use of WS Atkins' specific expertise in waste minimization;
- utilize Associated Octel's detailed knowledge of its plant and processes.

It was recognized that for waste minimization to be successful within Associated Octel in the longer term, there would need to be a transfer of waste minimization skills to Associated Octel personnel.

With these factors in mind it was decided to set up a site water team comprising representatives of:

- Health and Safety & Environmental Affairs (chairman);
- Site Manufacturing Groups;
- Site Effluent Treatment Plant;
- Engineering Group;
- WS Atkins (consultants).

This team had the overall responsibility for managing and implementing the project. Progress was reported to, and where necessary approval sought from, the senior management by the water team chairman at the Ellesmere Port site's monthly Environmental Committee meetings.

THE WATER AUDIT

Once top level management commitment has been obtained for such a project, the project team assembled and the scope and objectives defined, the next stage of a systematic waste minimization project is the waste audit. The purpose of the audit is to identify and quantify all the sources of usage and/or loss of the waste, in this case water. Again, great accuracy is not required at this stage. The main objective is to be able to rank sources of waste in relation to one another. For example, at the Ellesmere Port site it was important to confirm that washing batch distillation kettles used 500,000 m^3 per year, as opposed to the original estimate of 200,000 m^3 per year, but it is not significant at this stage if the actual usage turns out to be 550,000 m^3 per year. If improvements are proposed for a waste source at a later date it might be necessary then to refine the estimate, particularly if capital investment will be required. Overall, the target for the audit should be to get a balance of incoming to outgoing flows to within $\pm 10\%$. This is normally sufficient to confirm that all significant sources have been identified and that the estimated quantities are reasonably accurate.

A key part of the waste audit for the Ellesmere Port site was a number of plant inspections with production and/or technical support personnel who were intimately familiar with operations in that area.

The main aim of these inspections was to identify the sources of water usage and/or loss and, if consumption data were not readily available, where those data could be obtained. The information gained from the plant inspections on sources of loss was cross-checked against other sources including the site water distribution drawings and the various IPC applications.

Quantitative data on water usage and/or loss were obtained from a variety of sources including:

- IPC applications;
- previous technical reports;
- measurement/production records (for example, total site water supply and effluent discharge);
- calculations (for example, cooling tower evaporative losses);
- anecdotal evidence;
- budget allocations;
- production levels (and water content of products);
- original equipment design data.

It took approximately three months to collect all the relevant data from around the site and to use it to develop a comprehensive site water balance. Table 1.3 gives a summary of the overall water balance. The sums of incoming water flows compared to the outgoing flows agreed to within 2%, indicating that all significant usages and losses had been identified and quantified reasonably accurately.

WHAT WAS LEARNED

Initially one of the most striking results was the realization that up to 40% of all the River Dee water entering the site was lost either in leakage from the distribution systems or in evaporation and drift from the site cooling towers. It was difficult to be certain about how this loss was split between leakage and cooling tower losses, as the methods for calculating the data were relatively simplistic. Also, this assumes that the metering of the incoming water flows is accurate.

In total these losses were costing over £600,000 per year. The sources could only be leakage, the cooling towers or inaccuracy in the meter for the incoming River Dee water, and therefore warranted further investigation. The

TABLE 1.3
Overall water balance — Ellesmere Port site

Inputs	m³ per year	Outputs	m³ per year
River Dee water	5,205,000	Effluent treatment plant —	
Towns water	265,000	liquid outfall	3,114,000
Rain water	305,000	Effluent treatment plant —	
Canal abstraction	6000	off-site disposal	1500
		Products leaving site	36,500
		Cooling tower losses	
		(evaporation and drift)	640,000
		Losses from main stack	39,000
		Drains direct to sewers	117,000
		Storm water to Manchester	
		Ship Canal	248,000
		Losses to ground	1,466,000
TOTAL	5,781,000	**TOTAL**	5,662,000

approach to be adopted is to contact North West Water to arrange an independent validation of the incoming water meter. Additional metering is being considered for key points around the site cooling and process water distribution system.

These actions would confirm the amount of leakage and hence the level of resource that will be appropriate to expend to minimize it. In the longer term, increased monitoring of water flows throughout the system will lead to improved management of water on site and in particular will enable any significant leaks to be identified and rectified sooner.

Most of the rest of the water brought onto the Ellesmere Port site is ultimately discharged to the Manchester Ship Canal from the site effluent treatment plant — that is, approximately 3,000,000 m³ per year. The majority of time in the water audit was spent identifying and quantifying the sources of this aqueous effluent. The primary objective was to identify which parts of the manufacturing processes on site made the most significant contribution to the total effluent discharge. Figure 1.1 (page 8) shows a simplification of the River Dee water flows around the site. Figure 1.2 (page 9) summarizes the top ten water users identified and quantified during the audit. If significant reductions in process water are to be achieved it is these areas that must be tackled.

7

WHAT NEXT?

The top ten water users can be split into two broad categories. The first is those users whose water consumption is largely determined by the design of the process and equipment and is not significantly affected by the actions of the process operators. An example of these is the Pease-Anthony (PA) venturi scrubbers used to remove particulates from the exhausts of process furnaces. The water supply to these scrubbers has been fixed at the scrubber design flow rate and is

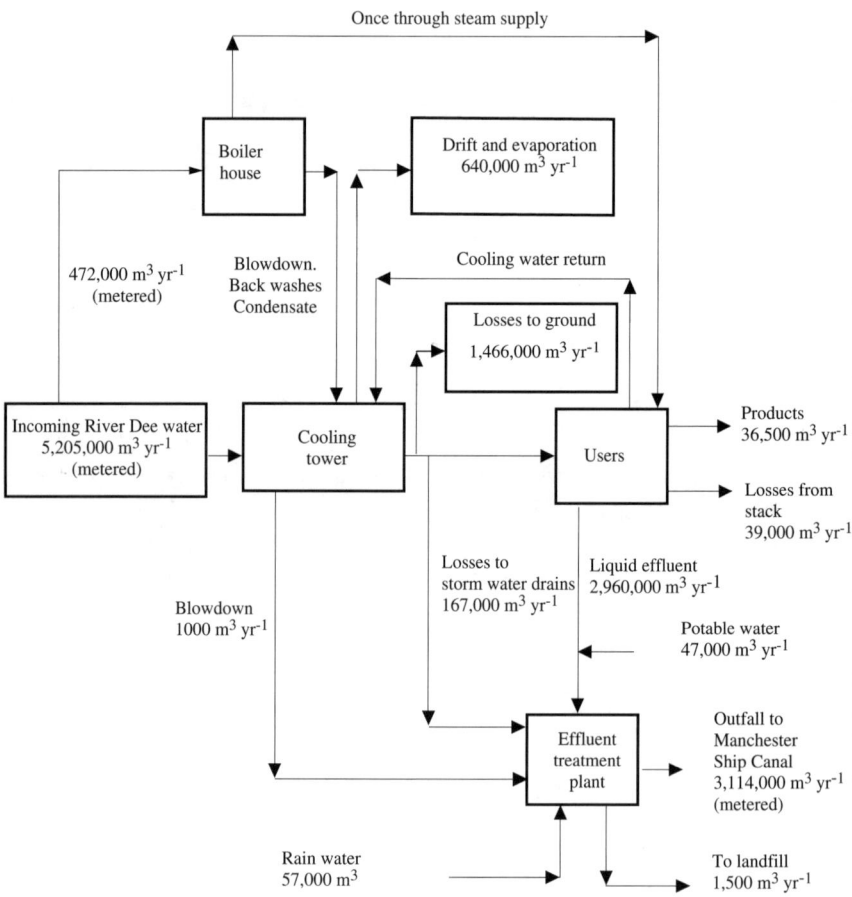

Figure 1.1   Simplified River Dee water distribution diagram.

not altered in normal operation. In the second category, water consumption is much more at the operator's discretion. For example, the amount of water used to wash distillation kettles in the MFAKC process is controlled manually by the process operators.

To take account of these differences, two approaches to generating the improvement opportunities will be adopted. Initially, the option generation will be undertaken in the three plant areas that use the most water.

In each of these areas a structured brainstorming meeting will generate as many ideas as possible to eliminate or minimize the largest users of water in that plant area, typically five sources of waste per session. Within the meeting all the ideas generated will be assessed for 'likelihood of technical success', their 'impact on waste if successful' and a note made of any significant capital cost implications. The output from these meetings will be a prioritized list of improvement options for immediate implementation or more detailed investigation and development. The teams for these sessions will comprise production managers, process technologists and process superintendents.

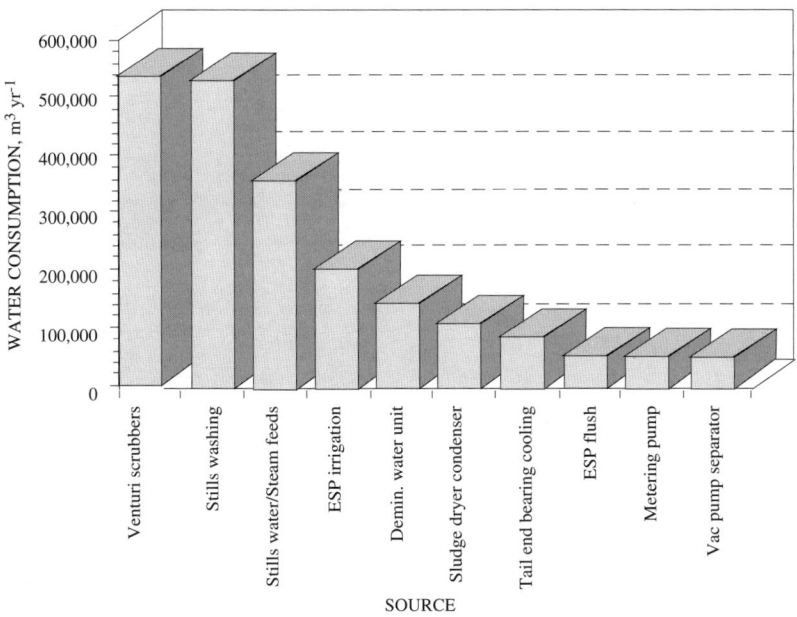

Figure 1.2   Top ten water users — Ellesmere Port site.

The options identified by the process outlined above will tend to be 'technical' solutions — that is, changes to the plant and equipment. Where consumption is largely controlled by the operators, changes to operating practices and procedures are often the best means of minimizing wastage. In these cases the solutions identified are usually better focused, and have a greater chance of being adopted in practice, if the ideas originate from the operators themselves.

Thus, a second set of brainstorming sessions is planned where the teams comprise process operators, maintenance personnel, process supervisors and process superintendents. The process superintendents are in the best position to understand the points of view of both the 'shop floor' and the 'management'; it is therefore important that they are involved in both approaches.

Although structured, the format of the brainstorming sessions for the operators is slightly looser. For the production management/technical support sessions the sources of waste to be addressed will be predetermined. In the operators' session they will be asked to identify sources of water use within their process area and, for each source identified, propose at least one idea for minimizing that usage. As well as utilizing the operators' detailed knowledge of their processes, to identify how best to minimize water use, this approach will also serve to raise the awareness of the operators to the environmental impact of their actions.

CONCLUSIONS

The water minimization project at Associated Octel's Ellesmere Port site has to date:

- identified that 40% of the site water usage is lost as leakage and from cooling towers;
- identified that 2,000,000 m^3 per year of aqueous effluent (out of a site total of 3,000,000 m^3 per year) arises from three plant areas;
- identified and quantified all the significant sources of water use and effluent generation on site;
- developed the strategy (implementation to commence summer 1996) to generate the minimization options in targeted areas of the process that will lead to significant water reductions.

Clearly this project is still far from complete. However, the findings to date demonstrate that significant reductions in water consumption are available and that significant savings in costs and effluent discharges will be achieved.

2. INTEGRATED POLLUTION CONTROL AT ALLIED COLLOIDS: THE 3ES CASE STUDY

Diana Cook

INTRODUCTION

Much of industry's perception of Integrated Pollution Control (IPC) is that stricter environmental legislation is more costly.

In November 1994, Dr David Slater — then Director and Chief Executive of Her Majesty's Inspectorate of Pollution (HMIP) — launched what has become known as the 3Es initiative. HMIP's aim was to work with industry to show, in David Slater's words, 'that IPC is not, overall, as expensive as some might have us think, and that it is, in fact, extremely good value for money'[1]. HMIP wanted to collaborate with a small number of companies and examine their whole experience under the heading of 3Es — emissions, efficiencies and economics.

It was recognized that Dr Slater's aim could not be achieved by simple retrospective analysis since the processes at Allied Colloids Limited had only been authorized under IPC since June 1994. Instead the joint project sought to develop and test a methodology for determining best available techniques not entailing excessive cost (BATNEEC) and assessing the impact of its application in terms of emissions, efficiencies and economics. Allied Colloids Limited volunteered to be the first industrial partner in this exercise and the joint project received the full backing of Allied Colloids Group Board.

The approach sought to identify the potential gains to be achieved by the application of IPC to specimen processes and to develop a methodology for identifying BATNEEC and quantifying its effects.

ALLIED COLLOIDS LIMITED

Allied Colloids Limited, Bradford, UK, is the main manufacturing site for the company's speciality chemicals. The company's products are chiefly high performance water-soluble polymers based on acrylic chemistry with principal markets in the paper, pollution control and mineral processing industries. In all of these principal market areas Allied Colloids products are used for environmental benefit — in waste water treatment and enhancing clean water recycling. Other markets for these products are in textiles, speciality coatings and the secondary recovery of oil.

The Bradford site has 18 IPC authorizations. IPC is supported on site by a specialist team who undertake technical environmental work and supply advice and interpretation on environmental legislation.

PROJECT SET-UP

Three study areas were chosen:
- an intermediates manufacturing plant — P199 plant;
- a semi-continuous batch unit manufacturing liquid dispersion polymers (LDP);
- the effluent treatment plant.

These areas were chosen because they were sufficiently different from each other and seen as reasonably efficient and therefore providing a true test for the methodology.

A cross-functional project team of Allied Colloids personnel and HMIP inspectors was set up. The team met monthly to review progress and plan the next steps. The technical work was carried out by specialist teams meeting as and when required.

There was a potential ambiguity of roles for the HMIP inspectors involved between the project work and their duties under environmental legislation. It was agreed at the first project meeting that, if at any stage HMIP inspectors needed to switch roles, then clear notification of this would be made. This did not cause any problems during the project.

The preliminary objectives were to:
- find out if the methodology worked in all three study areas;
- identify potential areas for improvements;
- evaluate 3Es process improvements against financial considerations.

The preliminary objectives were largely completed by August 1995, when an interim report was published[2]. Work since then has been on implementation of the improvements identified and a concluding report was published in March 1996[3].

PROJECT METHODOLOGY

HMIP North East Region had already developed an experimental methodology[4] which, it was felt, would act as a framework under which the 3Es could be examined. A comprehensive guide to the methodology was subsequently published[5].

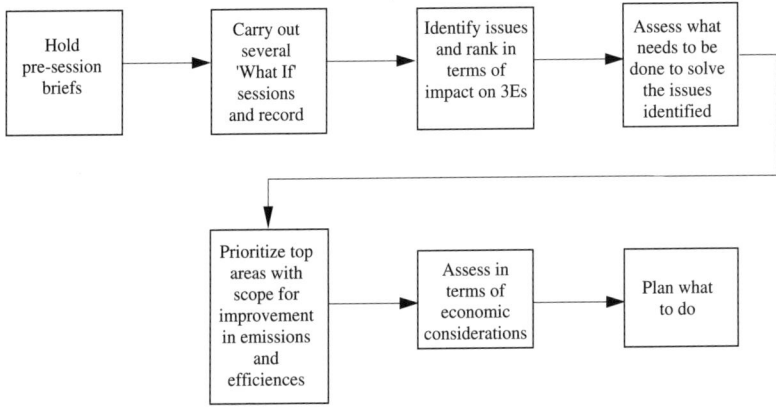

Figure 2.1   Flowchart for 3Es methodology.

The methodology applied the principles of IPC and process optimization through a structured review of the existing processes and work practices. Figure 2.1 shows the stages. These reviews took the form of 'what if' sessions. A list of guide words was used to focus a brainstorming on the existing processes and work practices. The aim was to establish a list of potential modifications which could benefit in an integrated manner all 3Es: emissions, efficiencies and economics.

The sessions worked best with a team drawn from a variety of disciplines and including not just experts on the study process but also those with a less detailed knowledge. The services of an independent chairperson familiar with running this type of session and able to impose discipline within the team was invaluable. It was essential at this stage that the team concentrated on identifying the problems, and did not try to solve them.

The next stages involved identifying issues from the mass of ideas, ranking these in terms of impact on 3Es and then assessing in outline what work needed to be done to solve the problems.

Once the economic implications of the emissions and efficiencies had been assessed, then the potential process improvements could be viewed in a balanced manner.

Various sources of existing company data were drawn on to help in the scheme of prioritization and ranking. Examples of available data were:
• mass balance;

- process effectiveness[6];
- plant maintenance records.

Even before the full implications of the study had been realized, participants felt that working through the 'what if' methodology generated a better understanding of the study processes. It gave environmental issues equal weighting with efficiencies and economics so that all three aspects could be considered in a holistic manner.

The participants also recognized that this project was not about implementing 'end-of-pipe' solutions or assessing what state-of-the-art technology was required. It was about fundamental questioning of the basic manufacturing process and work practices. It looked at finding process deficiencies and eliminating them, not recognizing symptoms and covering them up.

PROJECT FINDINGS

P199 PLANT

The P199 plant is situated in the middle of Allied Colloids internal supply chain (Figure 2.2). The upstream plant, P323, manufactures two products — dimethylaminoethyl acrylate (DMAEA) and dimethylaminoethyl methacrylate (DMAEMA). In P199 the products react with methyl chloride to give the two respective quaternized species.

The 'what if' study identified a number of situations whose occurrence would have significant environmental implications for primary and fugitive emissions. It also highlighted that plant efficiency issues such as storage and scheduling could influence primary emissions.

Three broad areas for improvement were identified:

- mass balance losses;
- potential emissions;
- plant efficiencies.

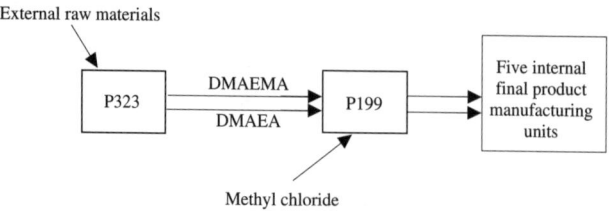

Figure 2.2   Intermediates manufacturing plant P199.

Loss of plant time was translated into financial terms by simply estimating the loss in profit from not manufacturing during these downtime periods. (This means of estimation relies on the assumption that the plant is at capacity and there is a continued demand for the product.)

It was estimated that total losses from these issues were about £1.5 million per annum.

Mass balance losses and potential emissions were tackled mainly through preventative system and engineering approaches. Breather valves on raw material or finished product storage tanks are now checked monthly by plant operators using Draeger tests to ensure that the valves are not venting outside defined limits. Since this programme was started, there have been no incidents involving failures.

Gasket failures on P199 can cause events ranging from emissions of odorous materials to spillages of reaction mixes at high temperature and pressure. The spillages would also result in the shutdown of other production units. A programme was started to ensure that all the gaskets on the plant which were in contact with the hot reaction mix were changed from compressed asbestos fibre (CAF) gaskets to PTFE-enveloped CAF gaskets. There have been no gasket failures of the new type.

The 3Es study challenged whether all flanges, gaskets, pipe routings and deadlegs were justified. Subsequently it was also found that the reaction mix attacks welds in the plant pipework. A programme was started to replace as many welded sections of pipework as possible with bends. Where welds could not be avoided, a high quality weld was used. As many flanges and elbows as possible were removed and T junctions simplified. Thicker piping with a longer life span was also used. At April 1996 about 20% of the replacement programme was complete and it was estimated that the remainder would be completed by the end of 1996.

This work contributes towards reducing both fugitive and primary emissions; it also reduces the risk of isolated events such as spillages.

On P199 'what if' issues relating to plant efficiency were prioritized and quantified by using the plant's existing process effectiveness database[6]; the number of times an efficiency related 'what if' question had actually occurred was obtained by reference to the number of times that same event had been recorded as a cause of plant downtime.

This showed that the two major events causing loss of plant availability were:

• control of inventory;

• scheduling.

Over a 21 week period these two issues accounted for 50% of the plant downtime and an estimated £180,000 lost profit per annum.

An 80-fold increase in methyl chloride emissions was measured during a product changeover — an event directly linked to inventory and scheduling systems.

The analysis indicated that this was a prime area to tackle both in terms of the opportunity to reduce the primary emission of an odorous and toxic material and in terms of lost profit and wasted raw materials.

The internal supply chain was 'mapped' in detail in terms of what operational management decisions were made and their cost implications. The exercise concentrated on the implications of the need to manufacture both products — if only one product was made there would be gains for all 3Es. On the basis of this analysis, it was decided to minimize manufacture of the minor product and source it externally where necessary.

Minimizing product changeovers contributes towards maximizing steady state operation of the plant. If other aspects of plant operation — for example, stock levels, production output, internal customer requirements — could run more smoothly, then this would also contribute towards minimizing emissions.

To help achieve this a simple spreadsheet model was built using the above parameters. The intention was that the model would be used by plant operators to help them maximize steady state running of the plant. Operators were trained both in the use of the model and in the 3Es philosophy behind it.

Evidence that changes have occurred as a result of the work carried out is shown in Figure 2.3. This data is derived from P199's process effectiveness database and compares two periods — the three quarters *before* and *after* March 1995. By the end of March 1995, the 'what if' study ranking and prioritizing had been completed on P199 and the first steps taken towards implementing improvements.

The main points are:

• plant downtime caused by storage and scheduling problems has been reduced by 33%;

• the number of product changeovers has been reduced by 55% and the number of associated lost plant hours by 80%;

• the number of incidents of mechanical breakdown have halved and the plant downtime associated with these issues reduced by 9%. Any problems with gaskets and breather valves are contained within this category.

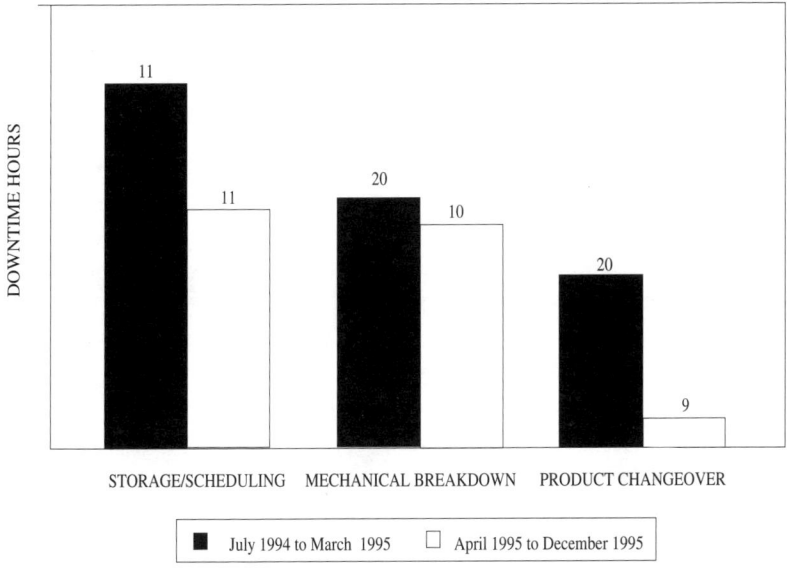

Figure 2.3   Comparison of plant downtime changes by category. Data labels refer to number of occurrences.

Figure 2.3 shows how reductions have been achieved in the number of mechanical breakdowns and product changeovers. These are examples of types of events which cause increased emissions. It is reasonable to conclude that methyl chloride emissions have been reduced, although it is not possible to quantify by what proportion.

Because all the work carried out so far contributes to several aspects of environmental improvement, quantifying the economic aspects of these improvements ideally requires a more sophisticated costing methodology than the 3Es project had available. However, it was estimated that around 25% of the originally estimated savings of £1.5 million per annum had been achieved.

Over a longer time period, the changes brought about by the improvement programme should be more readily quantifiable.

LIQUID DISPERSION POLYMERS
Liquid dispersion polymers (LDP) Division manufactures polymers using semi-continuous and batch processes to provide a wide range of products.

Figure 2.4 summarizes the manufacturing stages. Four reactor lines at the polymerization stage feed five driers.

The 'what if' study approached process optimization from the point of view of optimization of emissions (BATNEEC). Once the potential improvement in emissions had been identified, then the investigation of the underlying causes of those emissions indicated where efficiency and hence economic gains could be made.

Apart from some potential for reduced emissions of volatile organic compounds (VOCs), the main environmental improvement was related to secondary emissions resulting from the inefficient use of utilities. The nature of the materials handled by LDP meant that, unlike P199, primary emissions did not have the same immediacy of environmental impact.

The study showed that the principal cause of the emissions lay in imbalances at the reactor-drier interface. This resulted in a stop-start mode of operation rather than steady state.

The 'what if' study on P199 had already indicated that moving away from steady state operation increased emissions from P199. LDP confirmed these observations.

Figure 2.5 shows that for three out of five driers, waiting for feed from the reactors was the principal cause of plant downtime. When averaged over all the driers, this problem was the main cause of all drier downtime. The financial implications of this lost time were costed by estimating the loss in profit for an average LDP product from *not* manufacturing during these downtime periods. Figure 2.6 shows the results (see page 20).

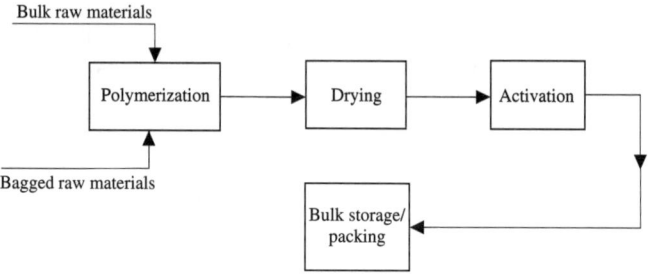

Figure 2.4   LDP manufacturing process.

The study concluded at this stage that if the imbalance between the re-actors and driers could be minimized then not only would secondary emissions be reduced but also that there were potential financial gains of around £3 million per annum. It was recognized that the data was crude; however, realizing even 10% of this figure per annum was very attractive.

Implementation of improvements centred around finding out what were the underlying causes of the reactor-drier interface problems.

Analysis of the output from a brainstorming session attended by plant management, operators and laboratory staff showed that the causes could be summarized by:

Production Links with Planning, People, Engineering and the QC Laboratory

Planning and people were ranked as the joint top priority issues with the power to effect the most change. It was felt that engineering and laboratory issues would have a lesser impact.

The solutions to the planning and people issues are mostly medium to long term and centre on the need to review the systems, responsibilities and lines of communication used to effect production planning.

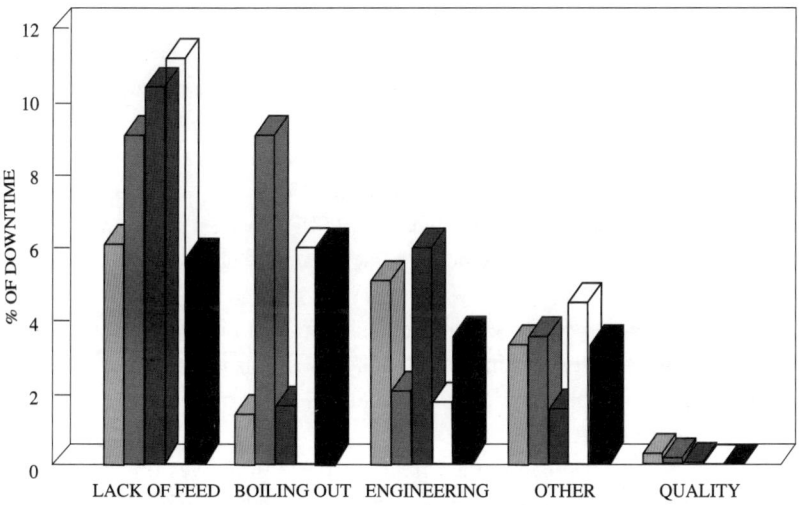

Figure 2.5   Causes of downtime on individual LDP driers, August 1994 to July 1995.

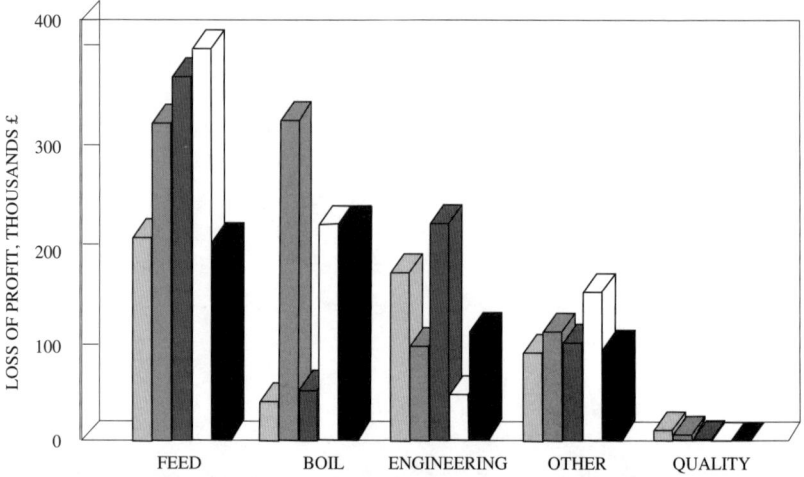

Figure 2.6   Lost profit from downtime on individual LDP driers, August 1994 to July 1995.

More short-term solutions are also being implemented. A simple spreadsheet scenario model is being built. The aim of the model is to help team leaders who work on different stages of the LDP process to see what effect their actions have on other parts of the process.

A particular aspect of the people issue concerns a need to review the roles and responsibilities of first line plant supervision; task analyses have already been carried out.

Investigations into how problems associated with the engineering aspects of the reactor-drier interface could be addressed were carried out by LDP's drier total productive maintenance (TPM)[6] team. This team looked at what items of equipment broke down most frequently and what could be done about them.

The laboratory concentrated on ensuring that production was not held up because of laboratory testing. The 'what if' sessions identified several areas where improvement here could have a beneficial effect on downtime and hence secondary emissions.

It is too soon to see the effects of these initiatives on reducing the downtime at the reactor-drier interface. Progress will be monitored and quantified by process effectiveness measures.

EFFLUENT TREATMENT PLANT

For process plants, relating losses from emissions and efficiencies to process economics requires estimates of their relationships to lost profit, losses in raw materials and so on. The effluent plant was more straightforward as the output from the plant directly related to environmental performance. The economics of any changes were more readily costed through standard disposal charges.

The effluent treatment plant is in compliance with the sewage undertaker's consent. The 3Es study was thus not an exercise in complying with 'end-of-pipe' legislation, but a structured study to apply the principles of BATNEEC to an effluent treatment plant.

Figure 2.7 illustrates the principle operating units of the effluent plant. There are two main treatment stages:

- removal of acrylic polymers using a dissolved air flotation (DAF) cell;
- removal of acrylic monomers by aerobic biodegradation.

The 'what if' study identified a range of improvement issues. The following are examples of medium- to long-term projects and more short-term issues which are being addressed.

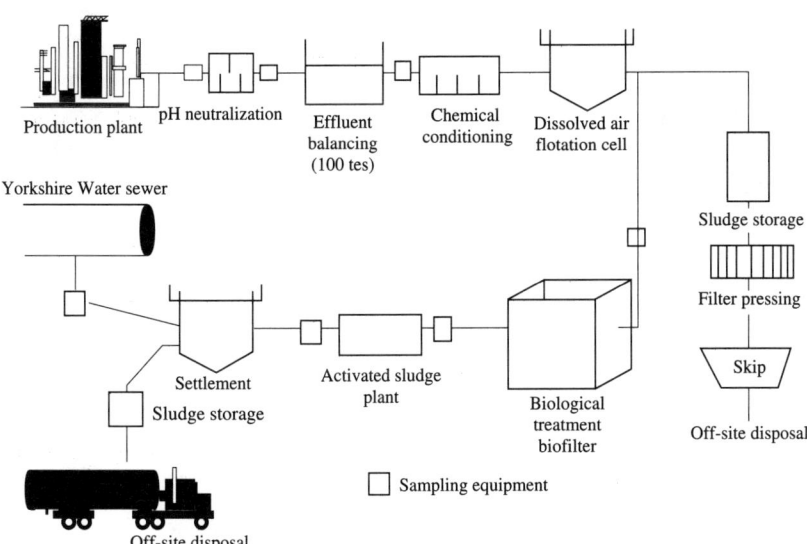

Figure 2.7   Effluent treatment plant.

An alternative method for effluent neutralization using carbon dioxide rather than hydrochloric acid was examined. Carbon dioxide would offer several environmental advantages over hydrochloric acid — better control over dosage, easier and safer to handle. Laboratory-scale work indicated that it would be cost-effective and a full-scale plant trial is now planned.

The 'what if' session highlighted the fact that, because of poor agitation, fresh sludge entering the DAF plant sludge holding tanks was channelled to the pump take-off point. This allowed old sludge to build up and led to several 3Es problems. Two new agitators have now been fitted to these tanks allowing a more consistent and drier press cake to be produced. The payback period is estimated at about two years, based on the fact that the presses can now operate more efficiently and special tank cleaning is no longer required.

Visual plant checks and analysis of process effectiveness data showed that throughput through the activated sludge stage was lower than expected. This was traced to the pump feeding the activated sludge unit. As well as problems with blockages, the pump was underrated. New pump designs have been investigated and an uprated pump will be installed in the near future.

As a direct result of the 3Es project, process effectiveness measures were set up on the effluent plant at an early stage. The measures are being used routinely to identify and quantify the effects of recurring plant problems. Improvements to the venturi on the DAF unit are one example of the beneficial use of process effectiveness information: blocked venturi caused frequent downtime. They have now been modified so that they are force fed with compressed air, removing the need for cleaning.

Visual plant checks revealed that throughput through the DAF unit was lower than it should be. The cause was traced to faulty non-return valves on the pumps feeding the DAF unit. These in turn caused effluent to by-pass the DAF unit — the part of the effluent plant where most of the COD load is removed. The repair work was quickly carried out at negligible cost.

The original estimates were that overall there was the potential for savings of £50,000 to £60,000 per annum in reduced disposal charges. Around two thirds of the potential has so far been implemented.

DISCUSSION

Compliance with environmental legislation is good business practice. The methodology gave a framework for going beyond compliance to a joint attempt to improve environmental performance. It was done not by imposing expensive hardware solutions but by emphasizing that an industrial concern running with high effectiveness and profitability will, by definition, have minimized its environmental effects.

The 3Es project has progressed beyond a list of issues towards the practical implementation of solutions to address the problems. The 3Es methodology is an approach to help establish BATNEEC.

It has been possible to demonstrate that work carried out on P199 and the effluent plant contributes towards reducing emissions — reduced primary and fugitive emissions to air from P199 and to land and water from the effluent plant.

Assessment of the financial returns of the work carried out to reduce emissions on P199 highlighted the complexity of environmental cost accounting. Conventional accounting practices do not deal adequately with environmental issues[7]. The techniques used here were simple and were sufficient to indicate where the priorities lay. There are now more detailed and thorough methodologies[7] available for evaluating the financial implications of large environmental projects. However, the use of their approach was beyond the scope of this project.

The early stages of the project were characterized by intensive effort from a small group of people. In the implementation stages, it was necessary to draw on a broader cross-section of Allied Colloids' expertise. This had several advantages; not only did it spread the work load, it also involved more people in the company appreciating and doing something about the environmental implications of their work.

Expenditure on engineering work so far is around £70,000, two thirds of this on the effluent plant. The indications from LDP are that there is a considerable amount to be gained from optimizing production systems, which will not require large capital expenditure.

CONCLUSION

The 3Es project at Allied Colloids has demonstrated that Dr Slater's original premise that IPC was 'extremely good value for money' has a sound basis.

Potential savings of about £4.5 million per annum were identified from the two production areas and £50,000–£60,000 from the effluent treatment plant. So far, one production area has been able to realize about 25% of the estimated potential savings in lost profit, and the effluent plant about two thirds in reduced costs.

Reduction in emissions to air, land and water has been achieved across the site.

The project has shown that the methodology was capable of identifying and ranking improvement issues on a variety of processes. The output from the methodology was useable; the issues were translated into action in the implementation projects. This process is a practical identification and implementation of BATNEEC. It also provides one of the tools to enable Allied Colloids to achieve its safety, health and environment (SHE) policy objectives.

Allied Colloids intends to continue the implementation and quantification of the work described here. The use of the 3Es methodology will be extended to other manufacturing plants utilizing in-house expertise.

REFERENCES IN CHAPTER 2

1. *ENDS Report*, 24 March 1995.
2. Allied Colloids and HMIP, August 1995, *3Es Project Interim Report* (available from both organizations).
3. Allied Colloids and HMIP, March 1996, *3Es Project Concluding Report* (available from both organizations).
4. Murray, M.P., April 1995, A review of environmental keywords in process/plant assessment techniques, *MSc Dissertation* (UMIST, UK).
5. HMIP and Business in the Environment, March 1996, *Profiting from Pollution Prevention: the 3Es Methodology*.
6. Nakajima, S., 1988, *Total Productive Maintenance* (Productivity Press, Cambridge, USA).
7. Moilanen, T. and Martin, C., 1996, *Financial Evaluation of Environmental Investments* (IChemE, Rugby, UK).

ADDRESSES

For further information on this project, please *write* to: Mr R.C. Barker/Mr R.C. McQuillan, Environment Agency, Stockdale House, 8 Victoria Road, Headingley, Leeds LS6 1PF, UK, or Dr D. Cook, Quality Improvement Manager, Allied Colloids Limited, PO Box 38, Low Moor, Bradford BD12 0JZ, UK.

3. SOLVENT MANAGEMENT — AVOID/REDUCE CAPITAL INVESTMENT IN VOC ABATEMENT*

Claire Shrewsbury and Nick Storer

INTRODUCTION

A Good Practice Guide (*GG13: Cost-effective Solvent Management*) has been published by the Environmental Technology Best Practice Programme (ETBPP) to help companies save money and lessen their environmental impact by the effective management of industrial solvents. An additional benefit of solvent management may be to delay, reduce or avoid the need to invest in volatile organic compound (VOC) abatement equipment, where such equipment may be required by regulations.

The ETBPP is a joint Department of Trade and Industry and Department of the Environment initiative managed by AEA Technology through the Energy Technology Support Unit (ETSU) and the National Environment Technology Centre. It promotes the use of better environmental practices that reduce business costs for UK industry and commerce.

For the purposes of this chapter, 'solvent' means an organic liquid that evaporates readily at normal temperature and pressure, giving rise to VOC emissions. The solvent is generally used or acts as a dissolver, dispersion medium or viscosity adjuster, or is used for cleaning operations. Companies using solvents operate in the following industrial sectors:
- surface cleaning and coating;
- printing and film coating;
- consumer product manufacture;
- paint and ink manufacture;
- seed oil extraction and food processing;
- adhesive manufacture and use;
- chemical manufacture;
- pharmaceutical manufacture.

---

*   © Crown Copyright 1996

Current levels of VOCs in the atmosphere are a subject of widespread concern and regulation, primarily because of their role in the formation of low-level ozone. Industrial solvents are a major source of VOC emissions in the UK.

This chapter explains how solvent management fits in with waste minimization schemes and environmental management systems such as those conforming to BS7750 and the EC Eco-Management and Audit Scheme (EMAS).

WHAT IS SOLVENT MANAGEMENT?

Solvent management is the process of gaining a better understanding of how a company uses solvents and how that company can control and reduce solvent consumption and the associated VOC emissions.

A *solvent management system* provides a structured and systematic way of achieving these aims. Solvent management can be either a stand-alone activity or part of a broader waste minimization programme, possibly within the framework of an overall environmental management system (EMS).

THE NEED FOR SOLVENT MANAGEMENT

Solvents have been used extensively by UK industry for many years. Despite the development of substitutes for some applications, there is still a need to use large quantities of solvents in a variety of industrial sectors.

It has been known for many years that some solvents and VOCs can harm human health. Many solvents are classed as hazardous; some are toxic or carcinogenic. The odorous nature of solvents and VOCs may also create problems, both on-site and in the immediate vicinity.

More recently, it has been recognized that some VOCs contribute to the destruction of the stratospheric ozone layer, act as global warming gases and take part in photochemical reactions with other substances in the air — particularly nitrogen oxides — to produce ground-level ozone which can potentially damage the human respiratory system, crops and natural vegetation.

These problems have brought the international community together to produce agreements on the control and reduction of use of VOCs. In the UK this has specifically led to legislative controls on the use of VOCs, such as:

- Control of Pollution (Special Waste) Regulations 1980;
- Environmental Protection Act (EPA) 1990;

- Environmental Protection (Prescribed Processes and Substances) Regulations 1991;
- Environmental Protection (Duty of Care) Regulations 1991;
- Control of Substances Hazardous to Health (COSHH) Regulations 1994.

THE BENEFITS OF SOLVENT MANAGEMENT

For many companies, the rising cost of solvents is as important as compliance with environmental legislation. For companies carrying out printing or coating operations, for example, direct solvent costs may be as much as 2% of their overall operating costs. Additional solvent costs are 'hidden' in the cost of inks, paints and so on. In other sectors, such as the pharmaceutical industry, solvent costs often represent 7–8% of operating costs. Since a tonne of solvent typically costs around £600 (1995 prices), even a relatively small percentage reduction yields worthwhile savings.

Solvent management can make companies more competitive and more profitable by reducing direct operating costs (through reduced solvent consumption, reduced consumption of associated materials, reduced disposal costs and possibly reduced insurance premiums), by improving efficiency and quality and by making a company more attractive to both potential and existing customers and investors. Solvent management can also eliminate the need for pollution abatement equipment by reducing solvent consumption below the registration threshold for Integrated Pollution Control/Local Authority Air Pollution Control (IPC/LAAPC) or reduce the cost of pollution abatement by process optimization — for example, reducing printing press airflows.

Increasing numbers of companies — including some smaller ones — are adopting corporate environmental policies and environmental management schemes (for example, BS7750), that require environmental 'vetting' of suppliers. Demonstrating a commitment to improved environmental performance could help to keep companies on 'approved supplier lists'.

THE SOLVENT MANAGEMENT FRAMEWORK

A solvent management system is a structured way of working to reduce solvent consumption and emissions. Figure 3.1 on page 28 shows a simple framework of six main steps.

Throughout the text and figures in this section, capital letters are used to indicate solvent flows through the whole site, and lower case letters are used

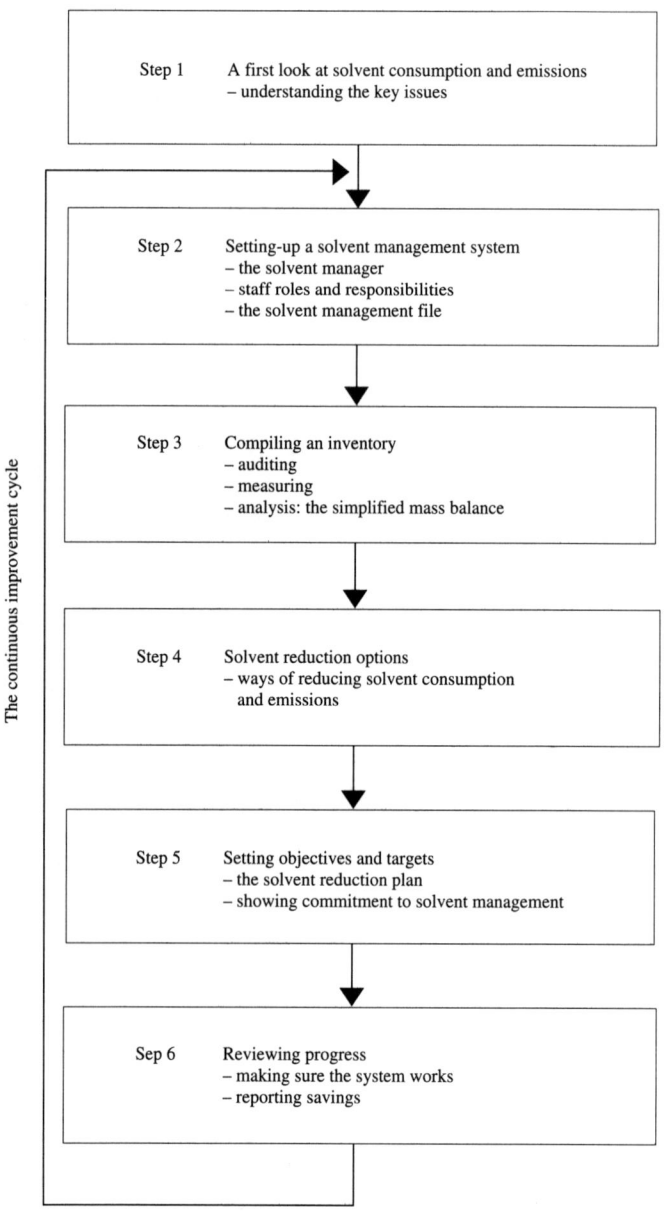

Figure 3.1   The solvent management framework.

when inputs, outputs and emissions are being considered for processes or stages of manufacture.

STEP 1 — A FIRST LOOK AT SOLVENT CONSUMPTION AND EMISSIONS
Commitment to solvent management from all employees will only be gained if it can be demonstrated that reduced solvent use will achieve significant financial savings and both the company and its employees will benefit.

As a first step, the technical manager, production manager or some other appropriate person should spend a few hours estimating how much solvent is consumed and how much this solvent consumption costs the company each year. Usually, the majority of solvent lost to the environment is emitted to air; it is therefore important to calculate these emissions.

How much solvent is emitted from the site?
The total amount of solvent emitted from the site (E) to the atmosphere can be estimated by carrying out a simple mass balance calculation. At first sight this may appear complicated, but by referring to the solvent flows shown in Figure 3.2, the equation can be quickly worked through.

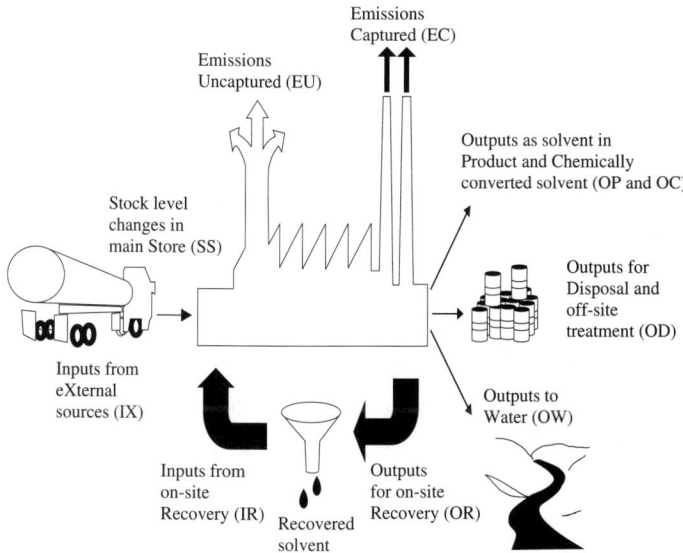

Figure 3.2   Solvent flows through a site or factory.

$$E = I - O - S$$

where:

$E =$ the total amount of solvent emitted to the atmosphere by evaporation during the year;

$I =$ the total amount of solvent purchased and/or reused during the year;

$O =$ the total amount of solvent disposed of, sent for recovery, reacted, etc, during the year;

$S =$ the change in stock level compared with that held at the beginning of the year.

It is important to use the same units for I, O and S (for example, $t \, yr^{-1}$).

Emissions to the atmosphere (E) consist of emissions that are captured (EC) — that is, extracted through ducting to a vent or stack — or emissions that are uncaptured (EU) — that is, released to the atmosphere via windows, doors and so on.

Inputs (I) include all solvents purchased or obtained external to the site (IX). This covers both virgin or recovered solvents and includes those present in coatings and solvents recovered on-site for reuse (IR) — for example, through distillation.

Outputs (O) include all solvents disposed of or treated off-site (for example, solvent and energy recovery, incineration, landfilling (OD)), solvents for on-site Recovery (OR) (for example, through distillation), solvent discharged to the sewer and/or a controlled Water (OW) (for example, a river, lake or coastal water), solvent left in the Product (OP) and solvent Chemically reacted in the process (OC).

Stock level changes (S) include stock changes at the main store and in the processes — for example, in machine reservoirs.

Different types of emissions to the atmosphere

Emissions to atmosphere (E), whether captured or uncaptured, can be intentional emissions from processes or machinery — for example, print drying — or unintentional or undesirable emissions — that is, losses from mixing and cleaning processes carried out either by hand or by machine, leaks, untreated spillages, open storage vessels and so on.

Determining emissions that are uncaptured

Once the total emissions to atmosphere (E) have been established, it is useful to determine the extent of emissions that are uncaptured. While some of the unintentional emissions — for example, from the mixing process — may be captured by extraction or ventilation systems, a large proportion may not.

Total Emissions to atmosphere (E) are the sum of the captured (EC) and uncaptured (EU) emissions:

$$E = EC + EU$$

Hence, emissions that are uncaptured (EU) can be estimated provided emissions that are captured (EC) and total emissions (E) are known.

Emissions that are captured (EC) can be determined by monitoring stack or vent concentrations and measuring stack or vent airflows.

For example, initial monitoring at a fictitious company might reveal that two extraction vents had, on average, a VOC concentration of 1120 mg m^{-3} and an air flow rate of 1.4 m^3 s^{-1} (5040 m^3 hr^{-1}). The site operates for 12 hr d^{-1} and 306 d yr^{-1}.

Emissions that were Captured (EC) were estimated as follows. Vent VOC concentrations were, on average, 0.00112 kg m^{-3}. Given a flow rate of 5040 m^3 hr^{-1}, each vent captured 5.65 kg hr^{-1}. The two vents therefore captured nearly 41,500 kg yr^{-1} or 41.5 t yr^{-1}. The company had already calculated that:

$$E = I - O - S = 66.6 - 8.25 - (-0.25) = 58.6 \text{ t yr}^{-1}$$

Emissions that were uncaptured (EU) are calculated as:

$$EU = E - EC = 58.6 - 41.5 = 17.1 \text{ t yr}^{-1}$$

If abatement equipment is fitted to a process, what is important for measuring emissions that are captured (EC) for solvent management is what goes into the stack or vent and not what comes out after abatement.

While such calculations are not precise (a small percentage error in I or EC could lead to a relatively large percentage error in EU), they establish an 'order of magnitude' for solvent losses through unintentional or undesirable emissions.

STEP 2 — SETTING UP A SOLVENT MANAGEMENT SYSTEM

Having highlighted the potential financial benefits of reducing solvent losses, the next step is to gain people's commitment to both the time and cost involved in setting up a solvent management system.

The solvent manager

It is important that an individual member of staff is given the responsibility of being the solvent manager. While the person responsible for health and safety issues is often the most appropriate choice, solvent management must not be seen as an additional burden.

It is important that solvent management responsibilities are clearly defined and written down in a solvent management file and that the solvent manager is made accountable both for the solvent management system and progress in reducing solvent consumption and emissions.

The solvent manager must be given sufficient resources in terms of time, people and equipment, and have the full support of employees at all levels.

The solvent manager's responsibilities include:
- making sure that the company complies with environmental legislation;
- making sure corporate and client environmental policies are followed;
- setting up the solvent management system;
- collecting and maintaining all records in the solvent management file;
- implementing a solvent reduction programme;
- helping to set realistic objectives and targets;
- reviewing progress and system effectiveness.

Specific aspects of the solvent management system are discussed at regular meetings, perhaps every two or three months, of the staff involved in health, safety and environment issues. Other managers, including the managing director, can be periodically involved in such discussions to keep them informed and committed to solvent management.

Solvent management file

This is simply a file containing all the relevant solvent management documents including:
- copies of up-to-date and appropriate extracts from all the relevant legislation, including Process Guidance Notes, COSHH, Duty of Care and Special Waste Regulations;
- all relevant corporate and customer policies, objectives and targets;
- a list of the solvent manager's responsibilities;
- complete records of all solvent audits and monitoring programmes.

STEP 3 — COMPILING AN INVENTORY

The solvent manager now takes a closer look at solvent use throughout the site or factory with the aim of identifying significant losses that can be targeted for reduction in the solvent management plan.

Making a better estimate of emissions from individual processes
In Step 1, emissions that are captured (EC) and uncaptured (EU) from the site as a whole were estimated. It is advisable to carry out such calculations regularly — for example, every one or two months. However, the solvent manager needs to know more about solvent consumption and emissions associated with individual processes (a process can be anything from an individual machine to a whole working area).

The solvent manager can consider the following actions:
* investigating the most likely problem areas by:
— carrying out fugitive or non-stack monitoring for leaks to identify areas with high VOC concentrations. This could be carried out simultaneously with COSHH monitoring;
— making regular physical inspections of storage tanks, drums, valves, pumps, flanges, seals, etc, visually, by smell and so on. For example, solvent leaks can sometimes cause ice to form as moisture in the air condenses with the cooling effect of solvent evaporation;
* developing a simple audit or inventory system that identifies solvent consumption and emissions for each stage within the manufacturing process.

Compiling an inventory: inputs, outputs and emissions from an individual process
The first stage in keeping an inventory of solvent consumption and emissions is to break down the manufacturing process into separate component processes, each with its inputs, outputs and emissions. Figure 3.3 on page 34 shows some possible solvent flows through an individual process.

Compiling an inventory: solvent flows through a typical small manufacturing site
Figure 3.4 (page 35) brings together five component processes — similar to the individual process depicted in Figure 3.3 — to illustrate the flow of solvents (not products) through a typical small manufacturing site.

Two processes can take place at the same physical location; for example, mixing and cleaning can be carried out at the side of a production machine. However, it is useful to keep such processes separate to help clarify the various consumption and emission figures.

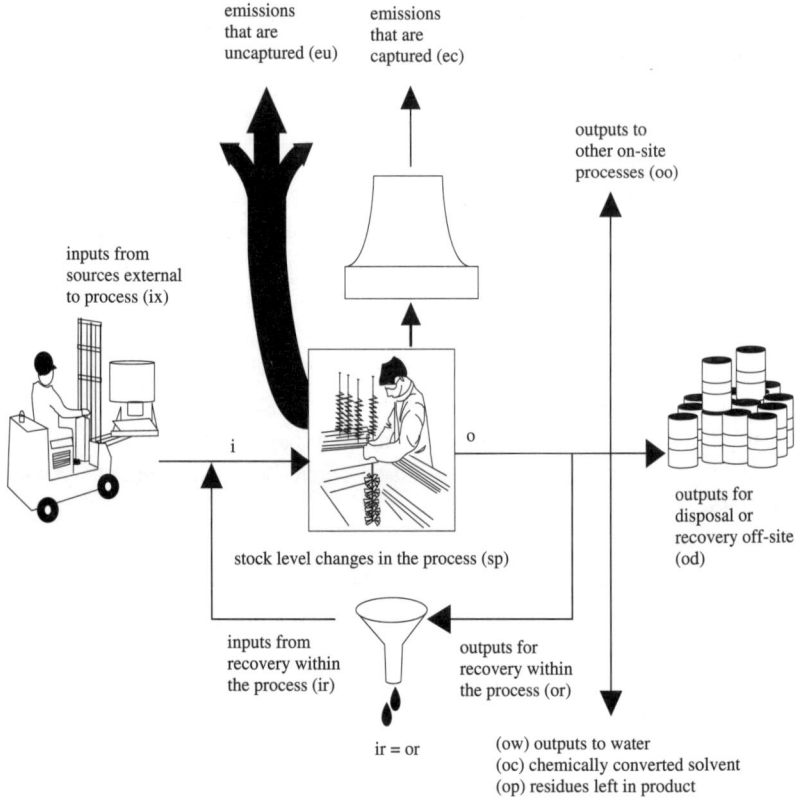

emissions
that are
uncaptured (eu)

emissions
that are
captured (ec)

outputs to
other on-site
processes (oo)

inputs from
sources external
to process (ix)

i

o

stock level changes in the process (sp)

outputs for
disposal or
recovery off-site
(od)

inputs from
recovery within
the process (ir)

outputs for
recovery within
the process (or)

ir = or

(ow) outputs to water
(oc) chemically converted solvent
(op) residues left in product

Figure 3.3   Solvent flows through an individual process.

Figure 3.4 shows an example of solvent flows that may need to be considered for a manufacturing site. While Figure 3.4 represents a slightly complicated manufacturing process, it is intended to highlight the links and components that a solvent manager may need to consider.

Essential record-keeping

Regardless of the layout and complexity of a site, it is useful to keep records relating to the overall site and specific processes. The solvent manager should ensure that the following records are maintained:

● site purchase records — a centralized record of all solvents and solvent-based coatings and products that are purchased or obtained off-site (IX);

• site dispense and transfer log — a centralized record of solvents and solvent-based coatings transferred from store (tanks/drums/cans) to a process (ix);

• process output log — outputs for disposal off-site from each process operation (od) are recorded individually by each process operator; outputs to other processes (oo) are also recorded. Any 'returns' of unused solvent to stores (not shown in Figure 3.4) are recorded centrally in the stores as part of the site dispense and transfer log;

• site recovery log — where on-site recovery is carried out; the operator of the recovery process keeps a separate record of recovery process inputs and outputs. Outputs from the recovery process are either outputs for disposal (od) or inputs for reuse (ir);

• site disposal log — outputs for off-site disposal (OD), including those from a waste water treatment plant, are recorded by the solvent manager or whoever is responsible for waste management. If separately notified special waste consignment notes are kept rather than annual blanket notifications, additional records may be unnecessary;

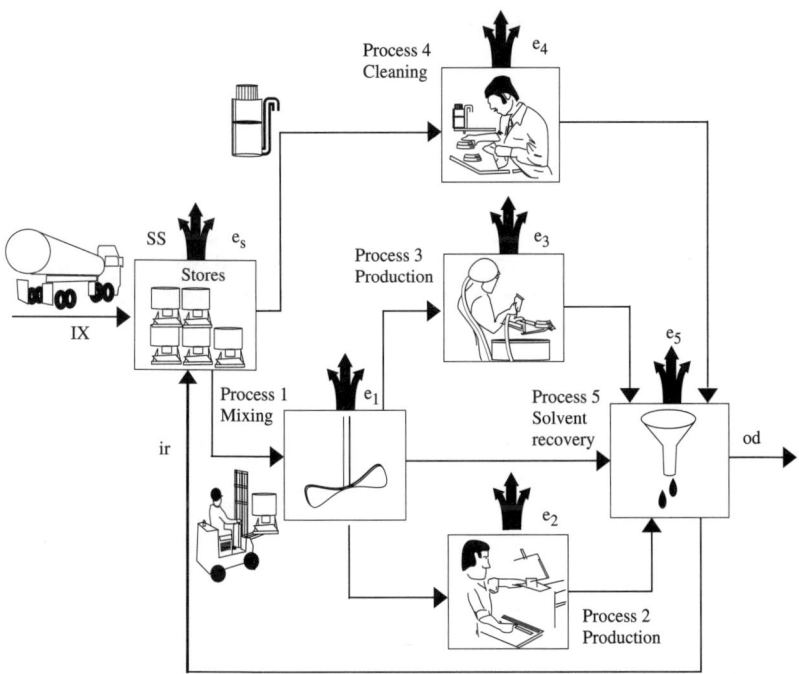

Figure 3.4   Example solvent flows through a small manufacturing site.

- spillage log — any significant solvent spills are recorded and reported to the solvent manager. Spills that are cleared up using absorbent materials — for example, rags — should appear on the site disposal log. Untreated spills will become captured or uncaptured emissions to the atmosphere.

In addition to providing useful audit information, reporting procedures for spills increases employee awareness of the need to be more careful when working with solvents. It is not necessary to record how much solvent is recycled within each process. Only *net* inputs and outputs are important.

Most companies already keep much of this information, although it may not be in the form required for solvent management. For example, solvent purchase and disposal records are needed to show compliance with environmental regulations. For many companies, implementing a solvent management system merely means modifying and/or supplementing existing recording systems and bringing data together.

Where several solvents are used, and particularly where solvent costs vary significantly, it may be advisable to keep track of individual solvents as they pass through the system.

This chapter describes an example of a solvent auditing system for a typical small manufacturing site. In practice it is necessary to modify the system to suit any particular site, but the basic elements and principles remain the same. It is important to remember that the more accurate the solvent auditing system, the easier it is to target and reduce solvent consumption and VOC emissions.

VOC monitoring

As with the site as a whole, emissions that are uncaptured (eu) from individual machines or processes can be estimated by calculating emissions that are captured (ec) on the basis of monitoring stack and vent concentrations. Some companies are required to monitor VOC concentrations regularly as a condition of an EPA authorization. However, the data are only useful in the mass balance process if monitoring is carried out upstream of, or without any, pollution abatement equipment.

Monitoring to determine emissions that are captured (ec) and hence to estimate emissions that are uncaptured (eu) can be useful and need only be performed periodically. Each company should decide how often it needs — and can afford — to carry out VOC monitoring.

Recording data

On a regular basis, and preferably every one or two months, the solvent manager collects together all the relevant log sheets in the solvent management file and then adds together all the relevant inputs and outputs to give total figures for the period in question.

Where additional outputs (ow, op and oc) are not negligible, these will have to be estimated separately. Generally they will not have been recorded in the process output log. It may be possible to estimate discharges to water (ow) by carrying out occasional experiments to capture discharges that would otherwise go to drain. Alternatively it may be possible to estimate ow from discharge consents and the composition of waste water treatment sludges. This information should appear in the site disposal log.

In addition, stock changes in the machine or process (sp) are noted for each process — for example, reservoir levels compared to the previous audit.

Calculating a simple mass balance for an individual process
Once the data have been brought together in the solvent management file, a simple mass balance can be calculated. The first step is to calculate process emissions (e = ec + eu).

$$e = i - o - sp$$

where:
e = solvent emissions to the atmosphere;
i = solvent inputs to the process;
o = solvent outputs from the process;
sp= stock level changes in process reservoirs.

This is exactly the same process as carried out at site level.

Thus in the example shown in Figure 3.4, the following can be identified:
- solvent emitted/consumed in mixing Process 1 (e_1);
- solvent emitted/consumed in production Processes 2 and 3 (e_2 and e_3);
- solvent emitted/consumed in cleaning Process 4 (e_4);
- solvent emitted from the stores (e_s) and recovery Process (e_5).

These figures have limited usefulness unless they can be linked to production throughput (t) over the same period of time — for example, one month. In the example in Figure 3.4, solvent use should therefore be adjusted to:
- solvent emitted per unit from Process 1 of total throughput = $e_1/(t_2 + t_3)$;
- solvent emitted per unit throughput from Process 2 = e_2/t_2;
- solvent emitted per unit throughput from Process 3 = e_3/t_3;
- solvent emitted per unit from Process 4 of total throughput = $e_4/(t_2 + t_3)$.

It can also be useful to relate process waste (od) (not shown on Figure 3.4) to throughput, for example comparing od_2/t_2 and od_3/t_3.

Production-adjusted figures allow the solvent manager to identify trends in solvent consumption both from month to month and from process to process. Leaks, poor working practices and so on thus become easier to detect.

However, the unit of production throughput must be chosen carefully to allow meaningful comparisons. This is not a problem when a site produces only one type of product. For example, a shoe manufacturer that makes one style of shoe can use pairs of shoes as a production indicator. Where a company produces a range of products, the amount of solvent used each month will vary. The solvent manager has the task of choosing the most appropriate production indicator for the site's particular production characteristics.

Determining uncaptured emissions (eu) from an individual process
Once the total emissions (e) associated with each process have been established and related to production throughput, emissions that are uncaptured (eu) can be determined. This figure provides an indication of unintentional or undesirable losses. Uncaptured emissions could occur in the input to the process (for example, leaks from a pumped system, spillages during transit from the storeroom, etc) or from the process itself (for example, leaks from the machine reservoir, ducting, etc).

Captured emissions must first be measured. Where these additional data are obtainable by monitoring, further mass balance calculations can be carried out for each process over the particular period of interest.

For Process 2 in Figure 3.4, for example:

$$eu_2 = e_2 - ec_2 = i_2 - o_2 - s_2 - ec_2$$

These values should be related to production throughput where possible, for example:
- captured emissions per unit throughput from Process 2 $= ec_2/t_2$;
- uncaptured emissions per unit throughput from Process 2 $= eu_2/t_2$.

This analysis process is illustrated in the worked example from another fictitious company.

This company uses two theoretically identical processes and items of equipment. Table 3.1 gives the data obtained for a typical one-month period.

TABLE 3.1

	Process 1	Process 2
Production throughput units (t)	1200	1040
Captured emissions (ec)	1800 kg	1520 kg
Captured emissions per unit of throughput (ec/t)	1.50 kg/unit	1.46 kg/unit
Uncaptured emissions (eu)	272 kg	269 kg
Uncaptured emissions per unit of throughput (eu/t)	1.23 kg/unit	0.26 kg/unit

While the two processes appear to have similar emissions per unit of production, the relative percentage values are quite different — that is:

- captured emissions from Process 2 are 3% less than Process 1;
- uncaptured emissions from Process 2 are 14% higher than Process 1.

A persistent difference in uncaptured emissions from the two processes indicates that a systematic, rather than random, solvent loss is occurring.

Other possible analyses

Two further activities may be worthwhile.

The first involves comparing the total site inputs (I) and the total stock dispensed (the sum of the various ix values), taking into account the Stock changes at the Store (SS). This will highlight any losses prior to dispense, identifying the combined effect of emissions to air from the store (e_S), discharges to water (ow_S) (for example, from any unrecorded spillages in the stores), and other losses (for example, theft of materials).

The second involves reconciling the combined process outputs for disposal (every od) with the total site Outputs for Disposal (OD).

Graphical analysis and presentation

Graphs and pie-charts are particularly useful when trying to convince other managers that steps should be taken to improve solvent management and can also be used to inform and motivate employees, providing a clear stimulus for co-operation and action.

Figure 3.5 provides an example of the type of diagram a solvent manager could use to demonstrate variations in solvent consumption between different machines, processes or even sites. Such graphs also help to identify trends and deviations from normal.

Changes in the consumption pattern shown in Figure 3.5 include:
- a seasonal variation, probably caused by summer temperatures increasing evaporation. Both processes are affected, Process 2 more so than Process 1;
- consistently greater solvent use by Process 2 compared to Process 1, even though they are same make, model and age;
- apparently more variable solvent consumption by Process 2 compared to Process 1 — for example, consumption in February and November is noticeably higher than expected.

Given this evidence, the solvent manager can consider possible explanations. It may be, for example, that Process 2 has an undetected leak or a particularly careless operator, while Process 1 has been better maintained or has a more conscientious and careful operator.

Converting solvent losses into cash terms provides a clear incentive for further action. Figure 3.6 indicates the differences between Processes 1 and 2 in terms of profit-loss per unit of production throughput.

Identifying and publicizing differences between processes, and in some cases sites, can be powerful in encouraging employees to adopt a positive approach to solvent management.

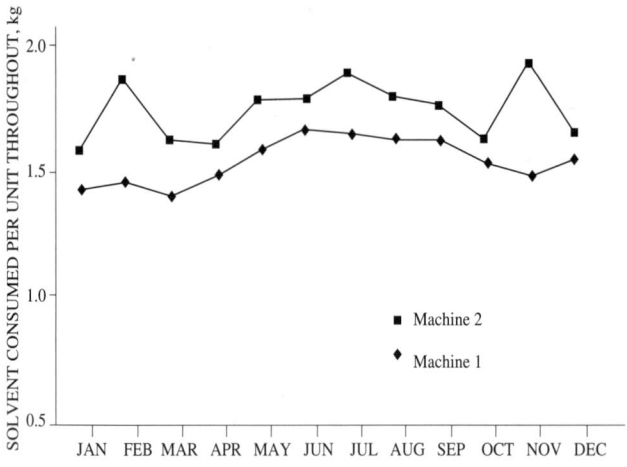

Figure 3.5   Solvent consumption variations with time at Moneywise Manufacturing.

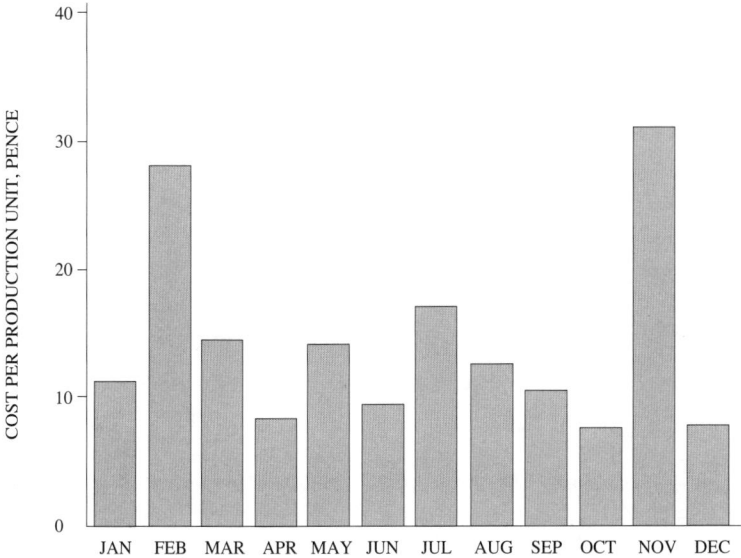

Figure 3.6   Financial losses per production unit for Process 2 at Moneywise Manufacturing.

STEP 4 — SOLVENT REDUCTION OPTIONS

The solvent manager can now use these data to convince other site or company managers of the need to take action to reduce solvent consumption and emissions. Once the company is committed to solvent management and the main areas of concern have been identified, the various solvent reduction options and their cost-effectiveness can be considered.

For those companies with more than one site, it is sometimes helpful to examine 'best practice' for identical or equivalent processes at the other sites. Solutions to specific problems may already exist within the same company.

Solvent reduction can be achieved in many ways. Target areas and approaches include:

• Purchasing and storage. Where possible, store in bulk tanks fitted with level meters and 'conservation' valves, paint tanks in light colours to minimize heating and evaporation, keep up to date with technological developments by evaluating compliant coatings and alternative cleaning agents, etc.

• On-site distribution, transfer and handling. Consider installing an accessible and easily viewed piped or pumped system to distribute solvents around the site

41

(for example, from sealed drums in the store to the mix and wash areas); where drums are used, provide adequate and safe trolleys to minimize the risk of spillage through drums being dropped; consider using intermediate bulk containers (IBCs) rather than drums, as IBCs can be carried more safely on forklift trucks.

- Process use. Consider eliminating processes where solvent use may not be essential, consider using low-solvent inks, compliant coatings, alternative cleaning agents or systems, consider installing condensation devices, consider process optimization (for example, reduced airflow rates can substantially reduce the cost of pollution abatement).
- Operator roles. Where solvent mixing is performed by different operators, consider training one person to carry out this task, preferably using an automated mixing machine. Train one person to be the main operator of the recovery equipment.
- Written procedures. Implement a set of procedures for each process that involves the handling or use of solvents. These procedures should emphasize ways of minimizing solvent loss through avoiding spillage, splashing, over-use, poor housekeeping, etc.
- Preventive maintenance and testing. Examine machinery, valves, flanges, pumps, tanks, etc, regularly to make sure that bolts are tight, etc, and that there are no obvious leaks. Pressure-test distribution pipework for leaks regularly.
- Staff training and motivation. Arrange regular informal seminars — such occasions can be a good way of establishing communication between management and plant operators. Link environmental awareness training to health and safety training. Publicize savings that have been achieved.
- Disposal and recovery. Consider investing in distillation equipment for on-site recovery of wash solvents and used coatings. Remove containers of solvent or coating residues from the shopfloor immediately to minimize the risk of spillage.

Selecting cost-effective solvent reduction options
Mass balance calculations, together with any monitoring and inspection activities, should highlight those approaches and solutions which are likely to reduce solvent consumption and emissions significantly. Some approaches have high capital, start-up and/or operating costs, while others only 'cost' a few hours of the solvent manager's time.

While no-cost and low-cost measures such as good housekeeping practices should be implemented immediately, it is also sensible to spend some time assessing the more expensive or difficult measures. The solvent manager

should, therefore, estimate the following for each solvent reduction option:
- likely reduction in operating costs through reduced solvent and associated materials purchases, reduced disposal costs and reduced insurance premiums;
- capital and start-up costs;
- additional labour costs.

Additional benefits should also be considered. These benefits, although sometimes difficult to quantify, include improved company efficiency, improved product quality and improved corporate environmental image — for example, winning or keeping contracts with environmentally-aware clients.

Each approach can be judged in terms of its 'value for money' by considering:
- the payback period on the capital or initial investment, taking into account any interest payments;
- the net yearly saving after the capital amount has been repaid;
- the risk of failure associated with each measure. Low-risk options with a relatively low payback may be preferable to high-risk options with a higher payback;
- the speed with which significant progress can be made. Tackling simpler problems first helps to increase staff confidence and motivation.

The various options can be ranked in order of preference by considering 'value for money', risk and the likely speed of progress. It may be possible to re-invest the savings from simple measures into high-capital solvent reduction projects.

STEP 5 — SETTING OBJECTIVES AND TARGETS

Once possible solvent reduction options have been evaluated and ranked, the solvent manager should discuss with the company manager or management team how much money the company can afford to spend on solvent management measures, particularly during the first year of such a programme. A choice needs to be made as to which solvent management measures are included in the annual plan. Finally, realistic targets are set for each chosen measure, and hence for the overall plan.

Targets are important because they provide a tangible goal that helps to motivate all members of staff. The overall solvent reduction target should be formally adopted as company policy and written into a solvent reduction plan.

The solvent manager can then make sure that all company staff are aware of the solvent reduction plan and its potential benefits in terms of greater profitability, job security and, possibly, financial bonuses.

The solvent manager directs the implementation of the solvent reduction plan. It is important that all employees, from the managing director down, are aware of, and involved in, the solvent reduction plan and its aims. The plan should not be 'imposed' on employees.

STEP 6 — REVIEWING PROGRESS

Action to reduce solvent consumption and emissions continues from year to year as part of a continuous improvement process. Each year, new measures can be introduced and new targets set.

At least once a year, the solvent manager and other relevant site managers should review the progress made in reducing solvent consumption and emissions. The review considers:

- the overall success of the solvent reduction plan in terms of a percentage reduction in solvent consumption and emissions compared to targets and net cost savings;
- the success of each measure in the solvent reduction programme (also in terms of reduced solvent consumption or emissions and savings);
- the reasons for any failures to save money and reduce solvent emissions;
- any difficulties experienced with the solvent management system;
- measures to be taken and targets to be set in the following year;
- ways of improving the solvent management system.

Results of the review should be widely publicized within the company. Regular feedback motivates employees and makes them feel more involved in the solvent management process.

ACHIEVING RESULTS WITH SOLVENT MANAGEMENT

This chapter describes a six-step, systematic solvent management system which is designed to help companies reduce both the amount of solvent they use and their VOC emissions. Ways to help improve the understanding of how a company uses solvents and where the greatest losses are occurring are suggested.

Although setting up a solvent management system takes time and money, it is important to remember that solvent management should increase a company's competitiveness. As a form of quality management, solvent management brings efficiency and product quality improvements. Through greater knowledge and control of solvent use, it may even be possible to reduce solvent and coating stocks — hence helping cash flow.

In some cases, reduced solvent consumption may mean that a company falls below the threshold for Part B processes and, as a result, no longer needs to obtain an authorization under LAAPC or to fit pollution abatement equipment. Alternatively for companies well within IPC/LAAPC, a reduction in solvent use and an associated reduction in emissions may lead to reducing the investment required in abatement equipment or at the least reduced running costs for existing abatement equipment.

Adopting solvent management principles assists companies in the preparation of information required under IPC/LAAPC authorizations. Solvent management makes a company more attractive both to 'green' companies further up the supply chain and to ethical investors. Being able to demonstrate that solvents are carefully controlled may mean reduced insurance premiums.

Start-up and operating costs need not be excessive. Most companies already possess records of the required data and hence only need to modify existing recording systems. Auditing and analysis activities need only take a few hours every month.

FURTHER READING

The following publications are available free of charge from the Environmental Technology Best Practice Programme.

- *GG12: Solvent Capture for Recovery and Re-use from Solvent-laden Gas Streams*;
- *GG13: Cost-effective Solvent Management*;
- *GG28: Good Housekeeping Measures for Solvents.*

For more information about the Programme or these publications, telephone the Environmental Helpline on 0800 585794.

4. TO BUND OR NOT TO BUND? A STUDY OF ALTERNATIVES TO SECONDARY CONTAINMENT[*]

Allen Ormond

INTRODUCTION

This chapter presents a case study. It shows how Eutech has provided a cost-effective approach to reducing spillage risks to Tioxide Europe Limited at its Grimsby site. It outlines the risk assessment approach, the nature of the improvement measures identified and how these were developed into an improvement strategy for the site. Eutech has provided a cost-effective approach to reducing spillage risks to Tioxide Europe Limited at its Grimsby site.

Like many operating companies handling potentially polluting chemicals, Tioxide Europe Limited (a subsidiary of ICI) was asked to undertake a cost/benefit analysis on improvements designed to prevent pollution from bulk storage areas on the Grimsby site as a condition of its IPC authorization. Bunding seemed an obvious improvement measure, but at very high cost.

Eutech's solution provided specific guidance on feasible and practical procedural improvements coupled with targeted improvements to selected plant equipment; the emphasis was on preventing spills rather than catching them. Other ICI plants have benefited from similar studies.

The approach is risk-based, initially involving a preliminary assessment of the types of incident that loss of containment might cause. The next stage was to model plant and operating data according to the more significant operational activities using a generic hazard assessment tool developed in ICI for rapidly assessing spillage potential.

Expert interpretation of the modelling focused on relevant improvement measures which could be readily compared in effectiveness with tank bunding. It was clearly demonstrated that:

- the alternative improvement measures provided risk reduction benefits that were similar to those of bunding, but would be far more economical. Indeed in some areas they would provide greater risk reduction benefits;

---

[*]   Copyright © 1996 Eutech Engineering Solutions Limited

• the added benefit of bunding beyond the improvement measures would be marginal and could be reasoned as excessive cost.

The approach has formed an integral part of the site's application for BS7750 registration.

BACKGROUND

The Environmental Protection Act 1990 (EPA 1990) requires many chemical manufacturers to submit applications to the Environment Agency, formerly Her Majesty's Inspectorate of Pollution (HMIP), for authorization under Integrated Pollution Control (IPC). Operators must demonstrate that they are making use of the best available techniques not entailing excessive cost ('BATNEEC') as a basis for drawing up improvement plans.

This case study is a practical example of how risk assessment of the potential for spillage from storage inventories at Tioxide Europe Limited's

Aerial view of Tioxide Europe Limited's Grimsby site.
© Geoffrey Pass Photography. Reproduced by permission of ICI.

Grimsby site has been used to demonstrate that a programme of less costly improvements renders tank bunding as only a marginal improvement measure and hence shows that it constitutes excessive cost.

Tioxide Europe Limited has been manufacturing titanium dioxide pigments at its Grimsby site on South Humberside for over 40 years. Manufacture is via the sulphuric acid route process. Concentrated sulphuric acid is manufactured at the site and distributed via a network of interlinked storage tanks. Caustic soda and other acidic liquors are handled within the make-up building, demineralization and de-alkylation areas of the site.

As part of the IPC authorization granted to the site in 1994, HMIP requested that a cost/benefit analysis be undertaken on improvements designed to prevent pollution from bulk storage areas. The concentrated sulphuric acid and caustic soda handling areas were recognized as those with the potential for serious pollution risk and were therefore selected for the study.

SCOPE AND BASIS OF ASSESSMENT

The assessment covered the concentrated sulphuric acid storage and distribution system along with the make-up building, demineralization and de-alkylation areas where the following chemicals are present:
- concentrated sulphuric acid;
- aluminium sulphate;
- caustic soda;
- caustic soda aluminate;
- titanium oxysulphate;
- trivalent titanium.

The more significant operational activities in these areas of the site were identified for assessment as in Tables 4.1 and 4.2 on page 50. Spillages were initially identified as having the potential to cause the following types of incident:
- excessive acid in the site's main effluent;
- ground contamination as a result of damage to storm drains;
- ground contamination by direct spills;
- discoloration of the site's main effluent by caustic soda precipitating suspended solids.

ICI Group guidance and judgement based on previous site experience was utilized to develop a benchmark for incident criteria. Two categories for defining the relative severity of incidents were defined:

TABLE 4.1
Operational activities assessed:
concentrated sulphuric acid storage and distribution system

1. Nos 4 and 6 plants production to tanks 5, 7 and 9

2. No 6 plant production only to tanks 5, 7 and 9

3. Tanks 5, 7 and 9 to east digester

4. Nos 3 and 5 plants production only to tanks 1, 2, 3, 4, 6 and 8

5. Nos 3, 4 and 5 plants production to tanks 1, 2, 3, 4, 6 and 8

6. Tanks 1, 2, 3, 4, 6 and 8 to tanks D and E

7. Tanker off-loading to tanks A and B during periods of on-site production difficulties

8. Tanks A, B, C, D and E inter-tank transfer

9. Tanks A, B, C, D and E to tanks 1, 2, 3, 4, 6 and 8

10. 98% cross-bleed on plant restart

TABLE 4.2
Operational activities assessed:
make-up building, demineralization and de-alkylation areas

1. 50% caustic off-loading and dilution to 22% caustic

2. Aluminium sulphate storage and transfer system into make-up building (MUB)

3. 22% caustic storage and distribution system

4. Caustic soda aluminate make-up, storage and transfer system to plants 4 and 5 areas

5. Concentrated sulphuric acid storage supply to anatase measure tank within MUB

6. Concentrated sulphuric acid storage supplies to demin plant and de-alk unit areas

7. Aluminium sulphate/titanium oxysulphate make-up

8. Trivalent titanium production, storage and transfer system to plant 3 area

- Category 1 — for a lower spillage quantity, generally with a tolerable frequency of once in ten years;
- Category 2 — for an order of magnitude higher spillage quantity, generally with a tolerable frequency of one in 100 years.

In addition, potential spillages to ground, that could occur for several hours because an operator is not regularly present to detect them, were considered more serious than more readily detectable spills. For these a lower tolerable frequency was applied.

Table 4.3 gives a summary of the incident types and associated tolerable risk criteria.

It is important to note that the tolerable criteria are used strictly as a benchmark for the assessment. Judgements over appropriate risk reduction measures are based entirely on the principle of 'as low as is reasonably practicable' (ALARP). Absolute levels of risk are therefore less significant than the relative reductions achievable.

TABLE 4.3
Incident types and associated tolerable risk criteria

	Incident Category 1	Incident Category 2
Tolerable frequency	$< 0.1 \ \text{yr}^{-1}$	$< 0.01 \ \text{yr}^{-1}$
Spillage quantity:		
Concentrated sulphuric acid to main effluent	$> 16 \ \text{Te hr}^{-1}$	$> 160 \ \text{Te hr}^{-1}$
Concentrated sulphuric acid to storm drains	$> 10 \ \text{Te hr}^{-1}$ Tolerable frequency $< 0.01 \ \text{yr}^{-1}$ if operator *not* regularly present	$> 100 \ \text{Te hr}^{-1}$ Tolerable frequency $< 0.001 \ \text{yr}^{-1}$ if operator *not* regularly present

Concentrated sulphuric acid direct to ground: as for storm drains.

Aluminium sulphate*	$> 54 \ \text{m}^3 \ \text{hr}^{-1}$	$> 540 \ \text{m}^3 \ \text{hr}^{-1}$
50% caustic soda*	$> 5 \ \text{Te hr}^{-1}$	$50 \ \text{Te hr}^{-1}$
22% caustic soda*	$> 11.4 \ \text{Te hr}^{-1}$	$114 \ \text{Te hr}^{-1}$
Caustic soda aluminate*	$> 9.3 \ \text{m}^3 \ \text{hr}^{-1}$	$> 93 \ \text{m}^3 \ \text{hr}^{-1}$
Titanium oxysulphate*	$> 14 \ \text{m}^3 \ \text{hr}^{-1}$	$> 140 \ \text{m}^3 \ \text{hr}^{-1}$
Trivalent titanium*	$> 45 \ \text{m}^3 \ \text{hr}^{-1}$	$> 450 \ \text{m}^3 \ \text{hr}^{-1}$

* All incidents are potential spills entering the main effluent system.

ICI's rapid spillage assessment tool was used to calculate the frequency of potentially contributing events to each incident category from plant and operational data.

ASSESSMENT METHOD

The rapid spillage assessment tool is a computer-based model. It conducts an assessment of generic means of loss of containment from storage installations considering operational failures/errors as well as failures of equipment integrity. It also estimates the magnitude of spillages/leaks so that they can be categorized according to the potential for severity. Mitigating measures from bunding/secondary containment can also be assessed.

Failure frequencies are defaulted by the model to generic data largely taken from in-house sources[1], but can be overridden by user input. An important feature of the assessment was the comparison of current risks predicted by the model with actual operating experience as reported by operating staff, so far as this was possible. In such a way plant and operating data were modelled and refined as necessary to improve the accuracy of the assessment in areas where it is critical.

The prime purpose of the tool is to focus on the main areas of concern and the most effective means of improvement.

The plant areas were modelled in some detail as storage and associated transfer systems, determined according to the operational activities involved. The results were collated via a common measure of the predicted frequencies of each event expressed as a percentage of its tolerable frequency. This provided a risk-based means of ranking the contributing events and hence prioritizing areas for improvement.

CURRENT RISKS AND REDUCTIONS ACHIEVABLE

Having established that the assessment of current risks calculated by the rapid spillage assessment tool was reasonably commensurate with known experience, a selection of risk reduction measures were identified for the highest risk contributors. Cost-effectiveness was of prime importance and emphasis was placed upon a priority ranking of:
- operating/procedural improvements so far as these could be feasibly and practically implemented;

TABLE 4.4
Risk (as % of tolerable criteria) improvements summary:
concentrated sulphuric acid storage and distribution system

Spills to:	Current	With procedural improvements	With additional equipment improvements	With bunding *only*	With improvements and bunding
Main effluent	21,300	2900	420	3900	100
Ground	230,000	35,700	920	46,600	410
Storm drains	52,200	13,800	460	13,700	240

$$\% \text{ of tolerable criteria} = \frac{\text{Predicted frequency of event}}{\text{Tolerable frequency for event}} \times 100\%$$
(summed for all events contributing to that incident type)

- equipment improvements;
- provision of secondary containment/bunding.

Table 4.4 gives a summary of the total risks in the absence of risk-reduction measures (expressed as a percentage of tolerable frequency), and the reduced levels achievable for the concentrated sulphuric acid storage and distribution system. Operating/procedural improvements here offer significant risk reduction benefits and when combined with selected equipment improvements provide a more effective benefit compared to the far more costly option of tank/pump bunding alone.

Realistically, the difference between the residual risks and the tolerable criteria are within the likely assessment accuracy. The benefits of bunding over and above those provided by the procedural and additional equipment improvements are not considered a worthwhile investment at this stage. The estimated costs of bunding are many times higher.

Tables 4.5 and 4.6 on page 54 summarize the current risks and the reduced level achievable by particular targeted improvements for the make-up building, demineralization and de-alkylation areas. Here the current risks are much lower, albeit tolerable risks are still exceeded. Tank/pump bunding would reduce the total risks to below the tolerable criteria. However, when compared to the achievable risks for the concentrated sulphuric acid storage and distribution system in the context of overall site risks, it is not considered a worthwhile

TABLE 4.5
Total spillage to main effluent risks table (% of category): current risks

%	Category 1	Categories 1 and 2	Category 2
System 2	611	664	53
System 3	0	0	0
System 5	713	713	0
System 6	1081	1081	0
System 7	0	62	62
System 8	0	500	500
System 9	0	0	0
System 10	52	52	0
Total	2457	3072	615

$$\text{Risk figures (\% of category)} = \frac{\text{Predicted frequency of event}}{\text{Tolerable frequency for event}} \times 100\%$$

(summed for all events in that system)

TABLE 4.6
Total spillage to main effluent risks table (% of category): after recommended improvements

%	Category 1	Categories 1 and 2	Category 2
System 2	44	97	53
System 3	0	0	0
System 5	115	115	0
System 6	72	72	0
System 7	0	62	62
System 8	0	10	10
System 9	0	0	0
System 10	17	17	0
Total	248	373	125

$$\text{Risk figures (\% of category)} = \frac{\text{Predicted frequency of event}}{\text{Tolerable frequency for event}} \times 100\%$$

(summed for all events in that system)

investment at this stage. The targeted improvements achieve residual risks similar to those achievable in the concentrated sulphuric acid storage and distribution system.

RECOMMENDED RISK REDUCTION MEASURES

Expert interpretation of the targeted improvements modelled led to the identification of specific improvement measures. To achieve the risk reduction benefits listed in Tables 4.4 and 4.6 the following[*] are being adopted by the site:

OPERATING/PROCEDURAL IMPROVEMENTS

(1) Full engineering inspection of all vessels in the concentrated sulphuric acid storage and distribution system targeted at remedial action to restore adequate vessel integrity.

(2) Condition monitoring of selected pumps:
- transfer pumps associated with concentrated sulphuric acid storage tanks A–E and 1–9;
- Nos 5 and 6 plants concentrated sulphuric acid storage product transfer pumps;
- 50% caustic off-loading pump;
- transfer pumps associated with caustic soda aluminate storage tank;
- titanium oxysulphate/trivalent titanium circulation pump.

(3) Audit existing procedures for tank decontamination/wash-out to ensure that discrete checks are made to avoid excess concentrated sulphuric acid storage put to drain.

(4) Audit existing procedures for operator checks on concentrated sulphuric acid storage tanks A–E receiving transfers, in order to:
- confirm, independently of the existing level measurement, that sufficient ullage exists prior to transfer;
- confirm that tank level changes are monitored as a 'route-check' during transfers.

---

[*]   Full details are provided here solely to illustrate how readily specific (rather than generic) improvements can be identified/targeted.

EQUIPMENT IMPROVEMENTS

(5) Install high level automatic shut-off systems to prevent overfilling of:
- concentrated sulphuric acid storage tanks A–E and 1–9;
- caustic soda aluminate make-up tank;
- caustic soda aluminate storage tank;
- dilute caustic pump tank;
- 50% caustic storage tank;
- concentrated sulphuric acid storage measure tank in de-alkylation area;
- 22% caustic storage tank in de-alkylation area;
- 22% caustic measure tank in demineralization area.

(6) Install a more reliable level measurement and high level automatic shut-off system to prevent overfilling of No 6 plant concentrated sulphuric acid circulation tanks. Additional protective measures against spillages to ground should also be considered.

(7) Install a backup seal on the 50% caustic off-loading pump.

(8) Route overflows direct to the main effluent system on concentrated sulphuric acid storage tanks A–E and 1–9.

(9) Route drain lines direct to the main effluent system on concentrated sulphuric acid storage tanks A–E, 6 and 8.

(10) Minimize flanges on the following pipelines:
- 98% concentrated sulphuric acid storage cross-bleed system;
- Nos 3–5 plants concentrated sulphuric acid storage product transfer from circulation tanks;
- concentrated sulphuric acid storage transfer line to east digesters;
- concentrated sulphuric acid storage recycle lines to storage tanks 5, 7 and 9;
- concentrated sulphuric acid storage product back line for No 5 plant;
- No 6 plant concentrated sulphuric acid storage feed to drying tower.

BENEFITS AND CONCLUSIONS
The main benefits to Tioxide Europe Limited of the study are:
(1) A comprehensive understanding of the critical aspects of the Grimsby site's storage and associated handling operations, sufficient to develop detailed improvement plans and evaluate the most cost-effective strategy. This enabled

56

clear decisions to be taken over whether or not the company should be committing itself to a bunding/secondary containment improvement programme and on which installations this would or would not be appropriate.

(2) Minimal cost and minimal involvement of plant resources when compared with the alternative approach to such an assessment involving attendance at hazard study meetings and detailed hazard assessments.

(3) Confidence in effective improvement plans. The assessment endorsed some of the plans already in place, and it also demonstrated their relative benefits/importance in order to aid prioritization. It also revealed some additional improvements that need to be considered.

(4) A risk-based assessment to enable objective discussion with regulatory authorities. The Environment Agency (formerly HMIP) encourages and increasingly seeks this from operators as a basis for their improvements; even their decisions not to improve. Experience has shown that in the absence of such evidence, regulatory authorities expect most storage/high inventory installations to be bunded and the costs involved tend to be much higher than alternative preventative/protective measures.

REFERENCES IN CHAPTER 4
1.   ICI Group Process SHE Guide No 14, *Reliability Data Manual.*

5. CLEAN DESIGN OF BATCH PROCESSES

Claire Houghton, Bev Sowerby and Barry Crittenden

INTRODUCTION

In recent years waste minimization and other environmental considerations have become more important to the design engineer, especially with the present Integrated Pollution Control (IPC) regulations. It is a natural extension that process designers will want to apply their traditional design methodologies to waste minimization studies and indeed much progress in this area has been made for continuous processes.

Generally, batch processes create more waste per tonne of product than continuous ones and different waste issues are dominant. Therefore, there is a need to create design methodologies that specifically address waste reduction in batch processes. A hierarchical methodology for the design and retrofit of batch and dynamic processes is presented in this chapter. Emphasis has been placed on analysing flowsheets at the conceptual design phase. In addition, dynamic simulation has been used to allow changes in critical parameters to be assessed quickly and to allow the effects of waste minimization efforts on the emissions from the plant to be examined.

HIERARCHICAL DESIGN METHODS

Hierarchical methods break down a complex design task into much smaller sections and deal with each in a strict sequential fashion. These methods allow a base case design to be developed logically starting at the simplest level and proceeding to the most complex. At each stage various options are considered and discarded if unfeasible in terms of operation or cost. This approach has several positive aspects: it helps the designer to understand the process quickly, it provides a current picture of the design economics, and it generates a list of process alternatives.

The first hierarchical design method for chemical processes was proposed by Douglas[1] who broke down the complex process synthesis task into the following decision levels:

Level 1: Batch versus continuous.

Level 2: Input-output structure of the flowsheet.

Level 3: Recycle structure of the flowsheet and reactor considerations.

Level 4: Separation system specifications.

    Level 4a: Vapour recovery system.

    Level 4b: Liquid recovery system.

Level 5: Heat exchanger network.

The entire process is considered at each level but as the design develops, further details are added to the flowsheet. Subsequently, Douglas adapted his original design methodology for waste minimization problems[2]. In his adaptation Douglas considered only the first four levels of design and took no account of energy recovery. This methodology, however, does not provide a full picture, as energy usage represents an environmental burden as well as an operating cost.

Rossiter and Spriggs[3,4] have taken Douglas's approach further and identified more general heuristics regarding waste minimization. The application of the Douglas hierarchical method to existing plants has also been examined by Fonyo et al [5] who used it to identify opportunities for retrofitting. Most of these studies concerned continuous processes. However, Rossiter and Spriggs[3] did consider batch processes.

An alternative hierarchical approach has been proposed by Linnhoff et al [6] for energy integration. Here an 'onion' diagram is used to represent the sequential or hierarchical nature of design, starting with the reactor and following with subsequent layers of separation system, heat exchange network and utility system. Smith and Petela[7] used the onion diagram for waste minimization by introducing waste considerations into each ring of the design. This approach allowed utility wastes to be included such as water and energy usage. The onion method is less rigid than the Douglas method but the philosophy is very similar.

The structure of the hierarchical design method described in this chapter has been modelled on Douglas's method. The structure, design levels, additional options and extensions are described below and several case study examples are used to show how the method can be applied to identify a variety of waste minimization options. The driving force behind this work has been to develop an easy-to-use screening tool. Emphasis has been placed on analysing flowsheets at the conceptual design phase and on using dynamic simulation both to assess changes in critical parameters and to allow the effects of waste minimization efforts on the plant wastes and emissions to be examined.

APPLICATION TO BATCH PROCESSES

For batch processes, the design process has been broken down into the following decision levels:

Level 1: Analysis of input-output structure.

Level 2: Analysis of reactor-recycle design.

Level 2a: Reactor charging.

Level 2b: Heating.

Level 2c: Reaction.

Level 2d: Vessel discharge.

Level 3: Analysis of separation system.

Level 4: Analysis of energy integration.

Level 5: Analysis of cleaning and scheduling.

A detailed description of these levels and the options to be considered at each level form the rest of this chapter.

LEVEL 1: ANALYSIS OF INPUT/OUTPUT STRUCTURE

The first level of this hierarchical design methodology provides a general overview of the process and helps to identify any waste streams that will be generated. The information required here is the amount and type of material input to the process, the main and side reactions that take place and the expected conversions and selectivities of the reactions. At this stage it is assumed that the principal reaction has been fixed and will not be altered unless a major waste problem is identified. Thus, an overall mass balance can be performed to determine the material outputs from the process. A basic heat balance can also be calculated from the heats of reaction to assess the amount of energy that will be required or emitted.

The design options to be considered at this level of the design are:

Are any waste problems caused by the reaction chemistry[2]?

A change to the reaction chemistry may not be an option to the process design engineer. However, if serious problems are foreseen it may be necessary to go back and reconsider the reaction chemistry.

Is spent catalyst a problem[7]?

Spent catalyst may be difficult to dispose of. Catalysts which can be regenerated may provide a better option. However, the regeneration operation may cause additional pollution problems. It might be better to use a catalyst with a long life

61

and to safeguard it within the reactor by ensuring that the correct flow, composition and thermal conditions are adhered to. Two examples in which spent catalyst causes a waste disposal problem are described in Case Study 1.

Can any additives or solvents be replaced to prevent waste?
For many batch processes, additives are required to inhibit, initiate or catalyse reactions. It may be possible to replace them with alternatives that generate less waste.

Are there any feed impurities that could cause waste problems[7]?
Impurities can cause problems by forming by-products, denaturing catalysts or affecting the main reaction efficiency by poisoning. Chemistry recipes for batch reactors are often very complex and it is possible that impurities in the feed could affect the reaction chemistry considerably. One solution could be to remove the impurity on site but this option might require additional energy input. Another solution could be to purchase purer feed stocks, but this would probably be more costly. Alternatively, it may be possible to remove the impurities before passing to a subsequent stage, although this could require additional separation stages.

Is there any potential for using waste streams in other batches?
In multi-batch plants a by-product may be useful in another part of the process. This may not be immediately obvious as batches may operate at different times and such opportunities need to be sought after. It may also be possible to recycle solvents and water to other parts of the plant.

CASE STUDY 1: INPUT/OUTPUT ANALYSIS — SPENT CATALYST
Many options exist and require consideration at the stage 1 design level for the treatment and disposal of catalysts. These two examples, taken from Integrated Pollution Control (IPC) applications, represent two alternatives.

Example 1
(Source of data: HMIP IPC Authorization No. AL 4716, Leeds[8])
Manufacturing of sodium cromolate uses a catalyst, TDA–1. After the reaction phase the catalyst is decanted to the on-site biotreatment plant with the effluent water. Sludge from the treatment plant, which contains the spent catalyst, is sent off site and disposed of in a licensed landfill site.

Total sludge sent to landfill = 2340 tonnes per annum

Total amount of catalyst used = 318 kg per annum

The amount of spent catalyst is only a small fraction of the waste sent to landfill. However, it does present a significant cost in terms of having to buy new catalyst. There may be additional landfill requirements due to the presence of the catalyst material and also an environmental cost of releasing a material to the environment.

Example 2

(Source of data: HMIP IPC Authorization No. AE 5772, Cardiff [9])

A platinum catalyst is used in the manufacture of 4ADPA (4-nitrodiphenylamine). Up to 0.05 kg of catalyst is used daily. Only small amounts are lost to the water effluent. After the reaction the materials are filtered and the catalyst removed. The catalyst is put in drums and sent to the suppliers for regenerating. The filter bags and materials contaminated by the catalyst are also sent to the suppliers for treatment and catalyst removal.

Regeneration of the catalyst may cause additional pollution problems. Could a catalyst with a longer life be used? Could the catalyst be safeguarded within the reactor?

LEVEL 2: ANALYSIS OF REACTOR-RECYCLE DESIGN

The second level in this design procedure concentrates on the reactor and recycle system. Here it is necessary to consider the different operational phases of batch processes as different environmental concerns occur in each phase. Typical operational phases have been taken to be reactor charging, heating, reaction and vessel discharge.

The initial stage of a batch operation is filling or charging the reaction vessel. Questions that should be considered at this stage of the design include:

Is the vessel open to atmosphere whilst charging?

If so, what and how much material will be emitted? These emissions may include residual material left in the vessel after the last batch or after cleaning with a solvent. If there is a problem then it is necessary to consider whether the vessel could be closed and filled under pressure. Case Study 2 shows how the emissions during vessel charging can be estimated using dynamic simulation, how these emissions can be reduced and what the consequences would be.

63

Does the order of addition affect the potential for releases?
In general the material with the lowest boiling point should be added first to dilute and/or dissolve any other materials that are added. Also materials that may react together and cause evolution of gas or a temperature rise should be added last so as to reduce the time available for emissions to take place and to provide thermal capacity in the vessel.

Does the method of addition affect the releases?
If a material is suddenly dumped into a vessel it may not enter solution quickly. Also if there is a high heat of solution the temperature rise within the vessel may become excessive. On the other hand, if the materials are metered in slowly the time over which emissions can occur is greater.

In batch processes, once a reaction vessel has been filled it is often heated in order to initiate the reaction. Options that need consideration here are summarized in Table 5.1. If the heating required is excessive then either the utilization of a catalyst in order to reduce the reaction activation energy or alternative reaction paths should be considered. The rate at which the reaction vessel is heated could affect the emission profile if the reactor is vented whilst heating. In addition, direct steam injection is often used as a method of heating in batch processes and this can result in a contaminated condensate. Condensate recycle could be an option although impurity build-up may then become problematic. Indirect steam, or electric heating via a jacket, could be viable alternatives.

Many pollution problems can originate from the reaction stage of the process and from the extent to which recycling is used. The following specific questions should be considered when the reaction chemistry is being analysed.

Should an excess of reactant be used[1]?
This would keep the concentration of reactant in the reactor high and in many cases would ensure a greater reaction rate and higher conversion. However,

TABLE 5.1
Reactor-recycle analysis — heating considerations

- How much heat is required to initiate reaction?
- From where is the energy obtained to achieve temperature rise?
- What method of heating is used?
- At what rate should the reactor contents be heated?
- What are the effects of heating during the reaction?

unreacted material would need to be separated and recycled; otherwise additional wastes would be generated. Excess reactant may also act as a diluent and therefore as a heat sink to prevent overheating.

Can the reaction equilibrium be shifted?
For systems limited by equilibrium the conversion can be raised in various ways depending on the stoichiometry[7]. Some reactions may be aided by the addition of an inert to the vessel. However, this can cause problems at the separation stage. Case Study 3 gives an example of the various factors that have been considered at this level for an esterification reaction.

Is complete conversion a viable option[2]?
Complete conversion would prevent the need for reactant separation, although this may require excessive residence times and energy inputs.

Can by-products be minimized?
Altering the reaction conditions could minimize the formation of by-products. This may also affect the main reaction and the desired conversion could become lower.

The next operational phase in the reactor-recycle analysis level is vessel discharge. This can lead to a variety of waste streams. The best way to deal with each output depends on the material, phase and destination. Treatment options need not be considered in detail but problem materials should be identified and avoided. If vapour phase waste products are generated they are usually vented or discharged to an end-of-pipe treatment facility which may involve recovery or destruction. Liquid wastes can either be discharged directly to drain, be recovered or be treated. The option of storing and using unreacted materials for future batches could also be considered. Waste solids are most likely to be sludged out with water or another solvent and sent to treatment or landfill. Options that avoid their formation should be considered at this stage.

Recycling options are the final considerations in level 2 of the design process. These could include the following:

Should streams be recycled?
For a batch system, material to be recycled might require intermediate storage vessels which could incur additional costs. Gas recycle could prove to be expensive due to gas storage and (re)compression.

At what point should recycled streams be returned to the system?
This may be restricted in some manufacturing sectors such as pharmaceuticals where backwards recycle is not allowable due to potential contamination of the product.

Should vapour recirculation (reflux) be used?
Normally, this is achieved by placing a condenser system after the reactor. Vapour recirculation can be used for pressure control by removing light vapours and for temperature control by recycling condensate.

Are any vents likely to release vapour during the reaction?
These releases need to be accounted for in the design as they will vary with time and may affect the impact they have on the environment. Concentration limits may be exceeded at a particular instant in time and might not be identified if only time-averaged effects were studied.

CASE STUDY 2: REACTOR-RECYCLE ANALYSIS
— EMISSIONS DURING CHARGING
To complement the qualitative design methods and questions certain operations have been simulated using the dynamic simulator HYSYS from Hyprotech. The example shown here is charging toluene into a 20 m³ vessel whilst the vent is open to atmosphere. The vessel is pressure rated to 10 bar g and the pressure rises to 4 bar g during subsequent operations. During this operation vapours are displaced from the vessel and contain traces of toluene. This vapour could also contain solvents used for cleaning the vessel or material left in the vessel after the last batch. The vent mass flow rate with time is shown in Figure 5.1.

The total mass of toluene released is 0.3 kg per batch.

If the vessel were closed during charging no emissions would occur. The pressure would rise and additional energy would be required to pump the material into the vessel.

Energy required = 11 kWh per batch

Cost (1996, South Western) = £2 per batch (approx)

The cost is small but energy consumption also has an impact on the environment. No additional equipment costs would be incurred as the vessel is adequately pressure rated. However, the rise in pressure may affect later reaction stages.

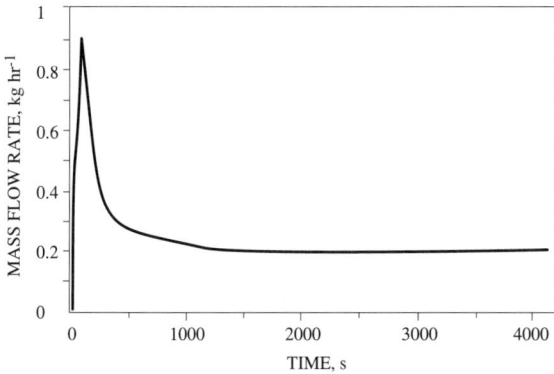

Figure 5.1   Mass vent flow rate of toluene versus time.

Another option is to vent the vapours back to the source vessel via a piped system which lets down the pressure. However, additional costs are associated with the installation of the pipes and valves required.

The challenge is to determine the best option for both economic and environmental viewpoints.

CASE STUDY 3: REACTOR-RECYCLE ANALYSIS
— REACTION CHEMISTRY OPTIONS
(Source: HMIP IPC Authorization No. 8044, Bedford[10])
There are many areas for minimizing wastes from batch reactions. This example is one of shifting the reaction equilibrium to increase the reaction yield.

Alkyds are produced by the following esterification reaction:

organic acid,	+	glycol	$\leftrightarrow$	polyester, alkyd	+	water
vegetable oil or		or				
fatty acid derivative		polyol				

Conversion can be increased by the removal of the polyester or water.

Alkyds are manufactured industrially using a solvent cooking process in which a small amount of Class B solvent is added and refluxed. As the reaction proceeds water distils over with the solvent and both are condensed into a separator. The water, contaminated with low levels of raw materials, is removed and disposed of off site. The solvent is recycled to the reactor.

TABLE 5.2
Separation system considerations

- Should the system be batch or continuous?
- What are the physical properties of the materials and how would these affect the system choice?
- Are additional materials such as solvents required?
- Is extra energy input required?
- Can new 'cleaner' technologies such as membranes be used?
- Can wastes be removed from effluent streams by further treatment?
- Are any streams the result of inappropriate or inadequate separations?

Solvent is vented to atmosphere during the resin production and can reach a flow rate of 0.0361 kg hr^{-1} for maximum operations. The presence of the solvent allows the temperature of the reaction to be decreased and reduces the oxidation of the products in air, so reducing the amount of nitrogen blanketing required as well as improving the yield.

The challenge is to determine which is the best option.

LEVEL 3: ANALYSIS OF THE SEPARATION SYSTEM

Table 5.2 gives a summary of the questions to be considered during the design of the separation system. Some of these have been taken from Douglas and Rossiter. Additional questions have been developed specifically for batch processes.

The separation process can be run directly from or within the batch reaction vessel or from a holding tank which acts as a buffer and allows the separation to be run continuously. A comparison between using batch and continuous separation methods, such as for distillation, should be made to see what effect these different operations have on emissions. Recently one company switched operations from seven batch stripping units to one continuous unit in order to reduce emissions[11].

Separation by adsorption, absorption and dissolving materials requires the addition of other mass separating agents in the form of solids, solvents, water and so on. Addition of these materials in themselves could produce more waste due to their ultimate disposal to landfill or losses to effluent or vents.

TABLE 5.3
Energy integration considerations

- Do any hot and cold streams occur at the same time in a batch cycle?
- Are there areas where there is high energy use or production?
- Is the efficiency of heat transfer limited by the temperature driving forces?
- How can energy consumption be reduced?
- Can the temperatures at which the heat is delivered be reduced?
- What fuels are used and are there any cleaner ones?
- Can the heat transfer coefficient be improved?

However, using distillation as an alternative to mass separating agents can incur extra energy usage. The advantages and disadvantages of these various options need to be considered.

LEVEL 4: ANALYSIS OF ENERGY INTEGRATION

Once a preliminary process flowsheet exists, the next stage in the design process is to identify all opportunities for energy integration. Pinch technology techniques have been used extensively for energy integration studies[12]. It is unlikely, however, that there will be enough information and time available at the preliminary design stage to carry out a full pinch analysis. The questions given in Table 5.3 have been developed to allow a simple energy study to be carried out. Such a simple analysis could be used to aid the reduction in energy usage, in terms of both cost and power consumption.

Douglas[2] did not consider energy in his waste minimization methodology. Rossiter[3] on the other hand did include energy considerations and his are the last four questions presented in Table 5.3. Energy integration of batch processes is complicated due to hot and cold streams existing at different times during the batch. Heat storage could be incorporated but this requires more equipment and a loss of flexibility. Therefore, although energy integration should remain an option during the conceptual design study, the complex analysis of energy storage and so on would normally be left to the detailed design stage.

TABLE 5.4
Cleaning and scheduling considerations

- What solvents are required?
- How many washes are required?
- Can the cleaning solvents be materials from within the process?
- Should the vessel be filled and then the solvent dumped? Or can surface cleaning techniques be applied?
- Could operation be counter-current multi-stage with a final rinse?
- Can a cleaning agent be used more than once or regenerated?

LEVEL 5: ANALYSIS OF CLEANING AND SCHEDULING

Consideration of cleaning and scheduling is not included in existing hierarchical methods which concentrate particularly on continuous processes. It is crucial in batch plants because significant amounts of waste can be generated in this phase. Key questions that have been developed for this level are shown in Table 5.4. The amount of cleaning required can be reduced by improving the production plans. Case Study 4 gives an example of a cleaning and scheduling problem.

CASE STUDY 4: CLEANING AND SCHEDULING OPTIONS
(Source: HMIP IPC Authorization No. 8044, Bedford[10])

For a multi-product plant cleaning can be a major source of waste. The amount of cleaning required can be reduced by optimizing the production schedule.

An example is a company making a variety of resins for the paint and printing industry. Two methods of cleaning are available:

- between batches of similar resins it is possible to clean the vessels with a solvent. The mixture of solvent and resin residues can be incorporated into future batches of similar material to avoid waste;
- if the products vary considerably a caustic wash is required which is sent to drain after settling of contaminants.

In this case the number of caustic soda washes is minimized in the production schedule by ensuring similar products are made in adjacent batches. This is limited by the customer demands and tight production schedules.

For process design other options could be considered such as the number and size of vessels.

CONCLUSIONS AND FUTURE WORK

A hierarchical design method has been presented that is based on Douglas' original method for waste minimization but with additional questions and levels that specifically address the waste problems of batch processes. Such a method will assist in identifying environmental problems during the design of each phase of a batch process and will give indications of how they might be overcome. Examples of waste minimization options have been cited and the use of dynamic simulation to calculate time dependent emissions has been presented. In future, more detailed dynamic models will to be used to test and further develop the proposed methodology.

REFERENCES IN CHAPTER 5

1. Douglas, J.M., 1985, *AIChE J*, 31 (3): 353–362.
2. Douglas, J.M., 1992, *Ind Eng Chem Res*, 31: 238–243.
3. Rossiter, A.P. and Spriggs, D.H., 1993, *Chem Eng Prog*, January, 30–36.
4. Spriggs, D.H., 1994, *Waste Management*, 14 (3–4): 215–229.
5. Fonyo, Z., Kurum, S. and Rippin, D.W.T., 1993, *Computers and Chem Eng*, S591–595.
6. Linnhoff, B., 1982, *User Guide on Process Integration for the Efficient Use of Energy* (IChemE, Rugby, UK).
7. Smith, R. and Petela, E., 1994, *Environmental Protection Bulletin*, Number 22, 3–10.
8. HMIP IPC Authorization No. AL4716, Leeds, Fisons Pharmaceuticals.
9. HMIP IPC Authorization No. AE5772, Cardiff, Monsanto.
10. HMIP IPC Authorization No. AJ 8044, Bedford, Croda Chemicals.
11. HMIP IPC Authorization No. AK 8708, Warrington, Elf Atochem.
12. Linnhoff, B., 1993, *Trans IChemE*, 71 (A5): 503–522.

ACKNOWLEDGEMENT

This research is funded by an EPSRC Process Engineering and Clean Technology Unit Research Grant GR/K27421.

6. MONITORING REQUIREMENTS UNDER INTEGRATED POLLUTION CONTROL*

Stuart Newstead and Robert Gemmill

INTRODUCTION

It is timely to discuss the present and future direction of monitoring requirements under Integrated Pollution Control (IPC). Compliance monitoring has
now become a requirement for most prescribed processes as the phased introduction of IPC nears completion. Furthermore, Her Majesty's Inspectorate of
Pollution (HMIP) has been incorporated into the Environment Agency since 1
April 1996. Consequently, HMIP's responsibilities are now being carried on by
the Environment Agency. This includes responsibility for regulating IPC prescribed processes in England and Wales.

HMIP established the National Centre for Regulatory Monitoring
(NCRM) as part of its preparations for merger into the Environment Agency.
The NCRM is carrying forward a number of initiatives to promote greater con-
fidence in IPC compliance monitoring data.

NATIONAL CENTRE FOR REGULATORY MONITORING (NCRM)

The role of the NCRM is to:
- develop monitoring strategies and procedures;
- carry out quality audits of monitoring arrangements;
- commission and manage contracts for national monitoring programmes and complex site investigations;
- provide related assessment, reports and guidance including indisputable evidence for regulatory action;
- promote technical excellence including the uptake of advanced measurement techniques;
- provide expert technical support on monitoring to Environment Agency staff.

Key objectives are ensuring quality and national consistency in compliance monitoring. These are recognized as being of particular concern to industry, and are brought into sharp focus by direct charging for compliance monitoring under the IPC Charging Scheme. The aim is to achieve these objectives through the development of appropriate monitoring strategies, the promotion of technical excellence and delivery of a cost-effective monitoring service.

MONITORING STRATEGY

The objectives of the NCRM's IPC monitoring strategy are to ensure that satisfactory arrangements are in place to:

• provide data on releases for assessments of compliance with limits in authorizations;

• provide information to the public in order to promote their confidence in the regulatory system;

• promote quality, reliability and improvements in compliance (and environmental) monitoring.

There are arguments for all monitoring related to regulation being carried out independently, by the regulator or its representatives. However, the

The interior of a mobile emissions monitoring laboratory — demonstrating instrumental analysers used by one of the Environment Agency's independent check monitoring contractors.

NCRM recognizes the growing complexity and sophistication of processes and requisite monitoring techniques, and that process operators are best placed to implement effective site-specific requirements. Given the scale of the task and the resources the Environment Agency would need to deploy, the favoured system remains for operators to carry out agreed monitoring of their own process releases, subject to Environment Agency audits and a proportionate amount of independent check monitoring to confirm results and provide public reassurance that the system is working effectively and honestly. This approach has the added advantage of requiring operators to take more of a first hand interest in the magnitude of releases from their processes.

OPERATOR MONITORING

Operators are required to use the best available techniques for monitoring. The Environment Agency regards these as on-line continuous measurement systems linked to computer data storage wherever practical, and not entailing excessive cost in relation to the process and its environmental impact. In many instances continuous monitoring systems are not yet available and discontinuous measurements, or testing, need to be used.

Under IPC, the operator may also be required to undertake environmental monitoring within the vicinity of a works. This measures the impact of releases on the local environment and provides information to the public on the consequences of permitted releases. It also gives evidence that the releases have been rendered harmless.

Operators are required to detail their proposals for compliance monitoring in their applications for authorization. Details should include:

• the locations on a plant, and in the environment local to the site, where monitoring is to be carried out. These include stacks, discharge pipes, areas resulting in fugitive emissions and key environmental indicators;

• the methods and frequency of monitoring to be used including equipment for on-line measurements and/or extractive sampling and laboratory analysis based on a BATNEEC (best available techniques not entailing excessive cost) approach. Preference should be given to continuous on-line real-time instrumental monitoring. When discontinuous testing is more appropriate, preference should be given to use of British Standard or international methods. In all cases it is for the operator to justify the proposed methods as 'fit for purpose';

• plans for deduction or calculation from feedstocks and an understanding of

Commencement of a test to BS3405: 1983.

process chemistry which may be an option where this can be shown to provide more reliable data;
- arrangements for recording and reporting data;
- quality assurance (QA)/quality control (QC) arrangements.

THE REGULATOR'S ROLE

Operators are required to provide results of their monitoring to the Environment Agency's staff for assessment and placing on registers where they can be examined by members of the public. In order to provide confirmation and reassurance that this system of self-monitoring is working honestly and effectively, operator arrangements are audited and a proportionate amount of independent monitoring commissioned by the Environment Agency undertaken.

Auditing is carried out as part of the Environment Agency's inspection duties and involves the physical assessment of operators' monitoring arrangements including the positioning, maintenance and calibration of instruments, sampling and analytical procedures and data recording and reporting arrangements.

Independent monitoring provides checks on operator compliance and the performance of continuous monitors, and supplementary information with which operators' data may be compared and the public reassured.

INDEPENDENT MONITORING
The Environment Agency's IPC independent monitoring is structured into three main sectors comprising:

ROUTINE MONITORING
Structured programmes of routine monitoring are undertaken in accordance with pre-determined schedules which are being developed on a process-specific basis. For example, separate programmes are drawn up for power stations, chemical waste incinerators, other combustion processes, waste disposal and other incinerators, minerals processes, metals industry processes and subsections of the chemicals industry.

AD HOC/REACTIVE MONITORING
In addition to the structured routine monitoring, ad hoc or reactive monitoring capabilities in the form of call-off contracts are in place. These provide flexibility to the Environment Agency's field staff to commission additional monitoring as part of their inspection duties, or to follow up unusual results from the routine programmes or as part of formal enforcement actions.

SITE SURVEYS AND INVESTIGATIONS
The routine and ad hoc/reactive monitoring can be supplemented by much more detailed site-specific surveys. These are one-off investigations extending from desk-top assessments of existing data to comprehensive environmental monitoring, and assessment of the environmental impact of releases using modelling techniques and by comparison with published standards.

RESULTS ASSESSMENT
Environment Agency staff assess monitoring data for compliance with relevant

release limits in authorizations or other criteria as appropriate. Comparisons are also made with previous returns to identify trends or unusual results and with the Environment Agency's independent data with a view to demonstrating consistency.

In the event that non-compliancies, abnormal results or gross inconsistencies are identified, the Environment Agency can take follow-up action including:

- discussing the data with the process operator to investigate the causes;
- requiring further operator and/or independent monitoring;
- varying the authorized limit where the present limit is regarded as unsatisfactory;
- enforcement action by improvement notice, prohibition notice, prosecution.

CONTINUOUS MONITORING INSTRUMENTS

The NCRM aims to promote and underpin the appropriate adoption of continuous monitoring instruments through publication of reviews of international practice in their application and use. These reviews have been, or are being, conducted by process sector. Publications available or planned are:

- *Continuous Monitoring Instrumentation for Emissions to Air from Large Combustion Plant*;
- *Calibration and Verification of Continuous Emission Monitoring Systems for Large Combustion Plant*;
- *An Assessment of Continuous Monitors for Incineration Processes* (joint publication with ETSU, the UK Energy Technology Support Unit);
- *Application of Continuous Monitoring Techniques to the Chemical Industry*;
- *Application of Continuous Monitoring Techniques to the Metals Industry*;
- *Assessment of Continuous Monitors for the Minerals Industry*.

DISCONTINUOUS TESTING

Past experience has shown that the standard of stack emission testing carried out by contractors, on behalf of both operators and the regulator, has been of variable quality. Typically, there has often been a lack of attention to the detailed requirements of sampling. Failure to sample processes representatively or at an appropriate point within a process cycle has been all too common. There have been frequent failures to adhere to standard procedures such as BS3405: 1983. This has sometimes been an inevitable consequence of poor sample point

Sampling of trace hydrocarbon emissions.

location, process fluctuations or the low concentrations at which the substances under measure are present. However, there are other occasions where contractors have shown poor perseverance, possibly because of commercial pressures. On certain other occasions, little appreciation has been evident of the need to note process conditions at the time of sampling. The lack of appreciation of regulatory requirements has been further demonstrated by instances of incomplete and sub-standard reporting.

Concern about variable and, in particular, indifferent contractor performance has been one driving force for monitoring organizations bringing about the formation of the Source Testing Association (STA). The increasing membership of this new body has been encouragingly rapid. The NCRM has supported the STA from its conception and continues to contribute to its discussions. NCRM experience of successfully improving the performance of the Environment Agency's contract monitoring teams is being used in work with the STA to develop appropriate performance standards for emissions testing. Issues under consideration include:

• the use and strict adherence to recognized standard methods as published by the British Standards Institution (BSI), Comité Européen de Normalisation (CEN) and International Standards Organization (ISO);

- the use of Environment Agency internal methods following validation through the Department of Trade and Industry's (DTI's) Valid Analytical Method (VAM) Programme;
- the use of Environment Agency approved measurement protocols;
- the use of standard reporting forms (requiring the entry of all relevant measurement details including key process information, so providing a traceable and auditable record from on-site recorded data to the final result);
- accreditation of all relevant on-site sampling and laboratory analytical methods with the UK Accreditation Service (UKAS, formerly NAMAS) as evidence of satisfactory QA procedures;
- certification and registration of stack testing personnel under a professional competency scheme preferably based on a BSI Standard.

MONITORING CERTIFICATION SCHEME

The Environment Agency regards continuous monitoring instruments to be the best available technology for demonstrating IPC compliance because of the greater temporal surveillance and more comprehensive information they can provide. The increasing interest in application of continuous monitors in the UK has been accompanied by an increasing lobby for a 'national' instrument certification/approval scheme. The main reasons for this are:

- performance standards and a type approval scheme are not presently available in the UK to assist industry in choosing monitoring systems which are fit for purpose, and promote public confidence in the regulatory regime's reliance on operator monitoring data;
- although schemes are established in some other countries, they have no official status in the UK. There is no internationally agreed scheme to follow;
- access to foreign schemes, particularly in Germany, has proved difficult, slow and expensive. The lack of instrument certification is believed to be placing manufacturers and suppliers of UK equipment at considerable competitive disadvantage in world markets.

With this in mind, HMIP, GAMBICA (the Association for the Instrumentation, Control and Automation Industry), and the DTI/DoE Joint Environmental Markets Unit (JEMU) funded a collaborative initiative to establish the basis for a UK certification service for certain categories of industrial stack emission monitoring instruments. A report was produced under contract by the National Physical Laboratory (NPL) and AEA Technology's National Environmental Technology Centre (NETCEN) proposing:

- instrument performance standards for determinands and process categories of most immediate interest;
- the possible structure of an EN 45011 based Monitoring Certification Scheme (MCERTS);
- recommendations for relevant testing (laboratory and field testing procedures) and post-certification procedures.

The report was drafted under the guidance of a steering committee and expert technical sub-groups made up of representatives from GAMBICA, HMIP, the Council of Gas Detection and Environmental Monitoring (COG-DEM), JEMU, DTI Environment and Energy Division, DoE Local Authority Unit, and the contractors themselves. The main information contained in the report was subsequently made available for public comment as an Environment Agency Consultation Document[1]. A second phase of work is under way to establish the scheme, including carrying out pilot studies.

It is still relatively early days for MCERTS and proposals are presently limited to continuous stack emission monitors intended for use on combustion, incineration and solvent-using processes. Performance standards are currently written for these specific applications and, more specifically, sulphur dioxide, particulate matter, oxides of nitrogen, carbon monoxide, carbon dioxide, oxygen, hydrogen chloride, volatile organic compounds, water vapour, temperature, pressure and stack flue gas mass flow monitoring instruments. However, it is recognized in setting up MCERTS that it has the potential for expansion, and provision is being made for incorporation of additional components of certification as and when required in the future. As well as development of the scheme to incorporate a wider range of determinands and processes for continuous stack emissions monitoring instruments, it is suggested that possible future components could include standards for indicative monitors, manual spot check monitors, ambient air monitors, water and effluent monitoring instrumentation, and for stack emission testing.

CONCLUSIONS

The Environment Agency is working towards a strategy of greater reliance on improved auditing of quality assured compliance monitoring regimes. This will be based on published performance standards to provide confidence that operators' monitoring data is reliable. The key to achieving this aim must be a commitment to professionalism, quality, reliability and accountability in compliance monitoring. The establishment of MCERTS is an important step in realizing this strategy.

REFERENCES IN CHAPTER 6

1. *The Establishment of a Monitoring Certification Scheme — Proposed Certification Scheme, Performance Standards, and Laboratory and Field Test Procedures for Continuous Stack Emission Monitoring Instruments*, Environment Agency's Consultation Report, April 1996.

7. THE ELECTRONIC MEASUREMENT OF ODOURS AND AROMAS

Chris Tullett

INTRODUCTION

The human system of smell has evolved over millions of years into a remarkable chemoreceptive system capable of detecting subtle changes in the air that we breathe.

Although impressive, our sense of smell can be unreliable — it is subject to emotional variables, physical changes such as colds and subjective influences such as the language used to describe an odour or aroma. Smell is therefore difficult to define objectively and, in order to help industry address aroma-related problems, a system is needed which recognizes aromas in a way that is similar to human perception but is also objective, accurate and repeatable.

Research in recent years has brought many advances in organic chemistry, electronics and computing which have made the electronic measurement and characterization of aromas possible. Developments by the University of Manchester Institute of Science and Technology (UMIST) and AromaScan have produced a commercial system which is helping to solve odour-related problems in the chemicals industry.

The system is based on the reversible interaction that takes place when volatile chemicals interact with the surface of conducting polymer sensors. The signals generated are used to form a 'fingerprint' of the aroma which can be compared with others and plotted in a two- or three-dimensional format. These patterns can be analysed using an artificial neural network, a powerful data processing tool. In this way the differences between a standard and test, acceptable and non-acceptable, or tainted and pure can be presented visually and with high levels of discrimination.

HUMAN MEASUREMENT OF ODOURS AND AROMAS

Despite extensive study the human olfactory system is perhaps the least well understood of all the senses. One thing is clear; it has a remarkable ability to distinguish a variety of odours. Whilst we can recognize and describe approximately 2000 different odours and aromas, the full potential of the human nose is considerably greater — a 'trained nose' such as a perfumist may be able to describe up to 10,000 different aromas, but only after years of training. The sensitivity of the nose is also quite remarkable. For example, one of the materials responsible for the musty smell in water — geosmin — has an olfactory threshold of approximately 5 ppt. Not everyone has this sensitivity and the normal human threshold varies between 5 and 200 ppt.

Smell is often regarded as an aesthetic sense, yet it is an important primal sense playing a vital role in interpreting our surroundings which may be essential to survival. It is this very important role that makes the human nose so weak as a system that is required to provide the sorts of consistent repeatable measurements over long periods of time that are required in a modern industrial environment. The human nose adapts to changes in its environment, screening out those smells that it knows are no longer a threat to the current situation, and in this way becomes temporarily blind to that smell after a few minutes of exposure. Everyone knows the effect of cold infections on sense of smell, but other influences — such as age, sex, and emotional state — also have an effect on the way smells are perceived.

Further variations occur in the way descriptions of odours are put together to describe a smell which may be interpreted differently by different people. The human threshold to volatile chemicals may vary considerably from person to person, so descriptions have to be carefully calibrated. Some volatiles — such as androstenone, a component of boar taint — cannot be picked up by significant sections of the population, yet females are known to have a significantly better sensitivity. Sensory panels therefore need to go to great lengths to standardize the descriptors used in this calibration process. This leads to a number of very important questions concerning the consistency of judgements made on odours using sensory evaluation. For precise odour and aroma evaluation, instrumental techniques are required.

INSTRUMENTAL MEASUREMENT OF ODOURS AND AROMAS

Any system that is required to measure odours and aromas in a practical commercial environment ideally needs to:

- emulate the human nose;
- measure odours and aromas objectively with high levels of repeatability;
- 'learn' from past experience;
- produce rapid results in a matter of minutes;

- be easy to use and interpret;
- be inexpensive to run.

Advances in organic chemistry, electronics and computing have made the electronic measurement and characterization of odours and aromas possible. Early work concentrated on tin oxide devices which operated at elevated temperatures of 300–600°C and relied upon the change in resistance as a consequence of gases being oxidized at the sensor surface. Recently a class of compounds known as conducting polymers, which operate at ambient temperatures, have been used in an array to characterize odours and aromas.

The use of arrays departs from the more traditional approach to chemical sensing where a specific sensor is manufactured for each analyte of interest. It is often difficult to achieve an interference-free signal when measuring complex odours and aromas using an analyte-specific sensor, so arrays are particularly useful when measuring complex headspaces. The resultant response or 'fingerprint' produced by a sensor array may show the response of certain sensors to be indicative of the general chemical nature of an analyte, but more importantly represents a response typical of that mixture of volatile chemical components that make up the headspace.

The AromaScan system combines the requirements of an electronic nose into a bench-top instrument that is helping to resolve odour- and aroma-related problems in a variety of different industries including food, beverages, chemicals, packaging, toiletries and environmental monitoring. The system uses a 32-sensor array of semiconducting polymers based on the electropolymerization of derivatives of pyrole, aniline and thiophenes which have been modified by attaching different functional groups to the ring (see Figure 7.1 on page 86). This ring substitution is unique to AromaScan and enables each element on the array to have a different conductive property.

When exposed to volatile chemicals, subtle molecular interactions with the sensor array produce changes in the conductivity of the polymer. Output from the array is simultaneously digitized and transmitted to a PC-based data system for storage, processing and analysis. Firstly the data can be represented as a bar chart showing the intensity of response for each sensor, known as the 'fingerprint' of the aroma (see Figure 7.2 on page 86).

When comparing several samples, there are too many data to handle at once to be practical, so data reduction is required. To observe these data graphically it is necessary to reduce them to two or three dimensions. The system utilizes Sammon mapping procedures to do this, and these show well how different samples compare in a given application (see Figure 7.3 on page 87).

Figure 7.1   32 conducting polymer sensors are used in the AromaScan sensor array.

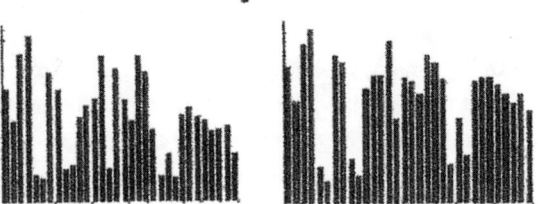

Figure 7.2   'Fingerprints' of two similar aromas.

Inevitably information is lost in the process and in order to utilize all the information an artificial neural network algorithm (fuzzified back-propagation) is used. By applying weighted factors to each of the 32 normalized sensor responses at two layers — input to hidden and hidden to output — the system generates descriptor nodes for each of the samples presented. This enables sub-descriptions within a descriptor to be defined. For example, at the output layer samples may be described globally (tea or coffee) with an associated sub-descriptor (Assam or Darjeeling). By using training data sets of known examples — for example, material with an acceptable level of taint and material with an unacceptable level — it is possible to determine product suitability objectively in a quality or process control situation. See Figure 7.4 on page 88.

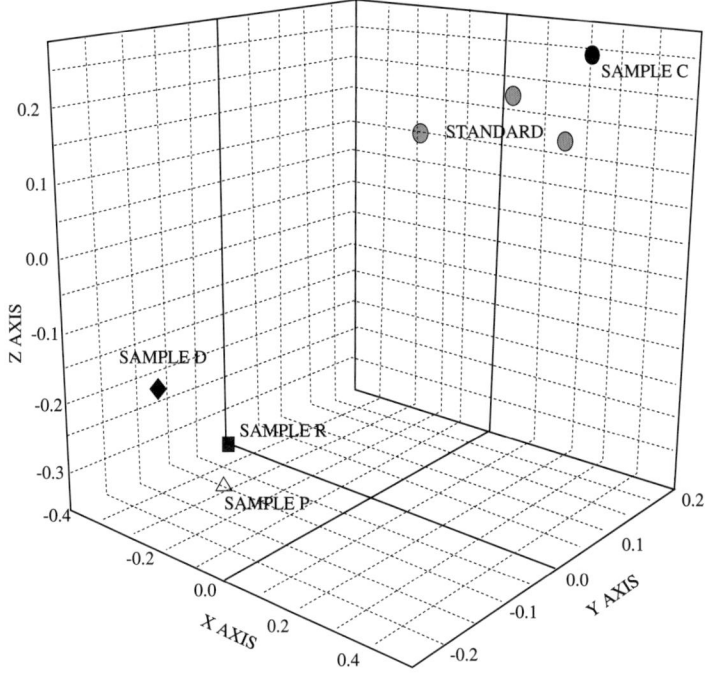

Figure 7.3   AromaMap showing discrimination between four different samples and standard.

An important consideration when measuring headspace is to ensure precise sample handling because the characteristics of an odour or aroma can be influenced by a change in conditions. The headspace must be allowed to develop under the same conditions of time, temperature and humidity. This is achieved with a sample pouch made of material that will not contribute to the overall odour and samples conditioned in a temperature-controlled cabinet. In the automated system samples are conditioned in glass vials closed by a septum sealed with a crimped cap.

APPLICATIONS
The technology has the potential for application in a wide range of environmentally important areas. These would include both on- and off-site monitoring and

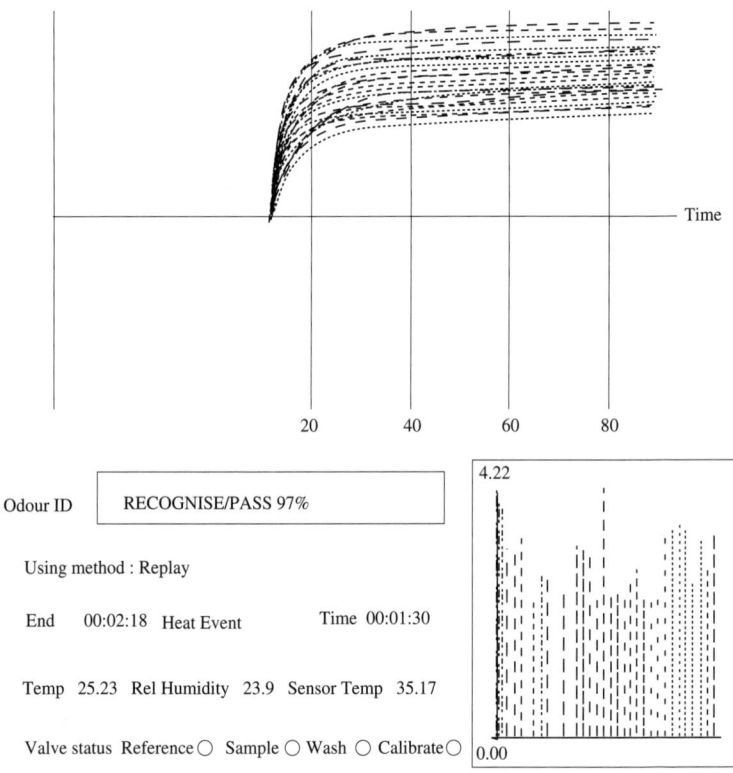

Figure 7.4   Data acquisition screen illustrating sensor response over time, fingerprint and artificial neural network pattern recognition.

control. The detection of odour in process vents, particularly where this would allow control action to avoid causing nuisance, could be very valuable. In situations where operators have odour nuisance problems control is made particularly difficult by the problems of measurement. Often, engineering changes need to be assessed in terms of a reduction of complaints from neighbours — a time-consuming process that is also likely to be subject to error and bias. The use of a rapidly responding, objective measurement technique has the prospect to allow a more effective response to odour pollution by operators. Off-site monitoring, particularly in industries that tend to have non-point odour releases (for example, food processing, rendering and waste water/sewage treatment) could also give early warning of problems and the opportunity to gain rapid feedback on engineering measures to control odour.

PLASTICS FOR BEVERAGES

Plastic materials are used extensively as packaging for a wide variety of purposes. Extruded plastic may be used as bottles which will eventually contain soft drinks. It may be used as a lining membrane for cardboard containers for food products. Many diagnostic vials are also made of plastic.

The ideal plastic packaging material would have no odour associated with it. However, manufacturing processes involved in the production of plastic do result in a significant amount of odour. The temperature at which the plastic is extruded can affect the odour of the final product. The odour associated with a batch of plastic determines whether it can be sold for a particular purpose.

If the aroma profiles of the plastic products or raw materials are assessed prior to leaving the factory and being shipped to the customer, the risk of rejection by the customer can be minimized. This type of quality control (QC) procedure offers significant cost benefits.

There are several ways in which the odour from plastic products can be analysed. These include sensory evaluation, olfactometers or gas chromatography/mass spectrometry; the analytical techniques are expensive, however, and do not necessarily reflect whether a measured odour is 'good' or 'bad'. It is undesirable to have the human nose exposed to the types of vapours given off by these products. Not only is it a subjective measurement but exposure to certain plastic emissions such as acetaldehyde can be harmful, causing a general narcotic reaction and respiratory paralysis in large doses. It is clearly advantageous to use an objective measurement system for these QC checks.

In this example, a user of plastic vials for diagnostic purposes needed an objective way to measure the odour of vials from different suppliers. Problems had arisen when high odour vials were used, causing incorrect diagnosis of some samples. The AromaScan system was used to evaluate plastic vials from two suppliers, as a quality control tool on the incoming product.

The AromaMap in Figure 7.5 on page 90 shows the data from plastic vials which were extruded at different temperatures (350°C, 360°C and 400°C). The distances between the points on the map are an indication of the odour associated with the vials. The vials extruded at 350°C and 360°C had similar aroma profiles and were acceptable. The 400°C vials had a pungent odour and were rejected by the customer.

BENZYL ACETATE FOR THE COSMETIC AND TOILETRIES INDUSTRIES

High grade benzyl acetate for the cosmetics industry commands a premium price up to 30% above low grade product. In this evaluation benzyl acetate was graded by a sensory panel and three different grades (samples 1 to 3) with replicates (samples 4 to 6) were measured by an AromaScan system. The system was trained using samples 1 to 3 and then asked to classify samples 4 to 6. As can be seen in Figure 7.6, the expected clustering was not observed. It is clear, however, that samples 4 to 6 belong to populations described by samples 2 and 3. This suggested an error in the labelling of the samples and extensive investigation by the customer found this to be the case. After re-examination it was found that the AromaScan system was classifying the grades of benzyl acetate correctly.

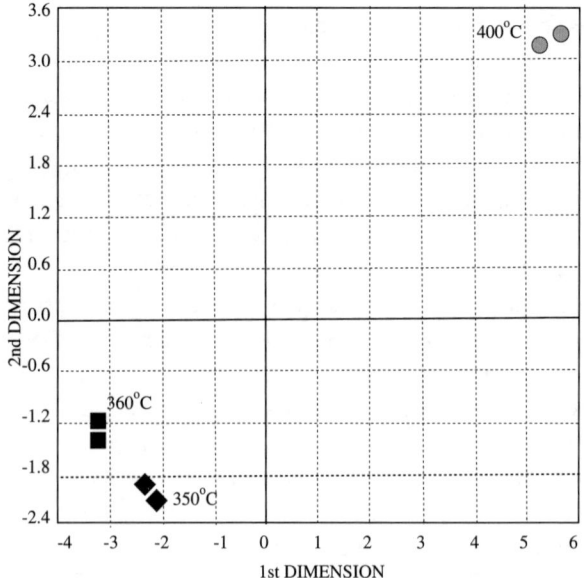

Figure 7.5   AromaMap showing the difference between the same material treated at different temperatures.

ELECTRICAL COMPONENTS

In the manufacture of electrical components the quality of isopropanol is important to achieve a satisfactory product performance. Differences in the purity of electrical grade isopropanol can have an important effect on the final product quality. An investigation using the AromaScan to discriminate between two grades was undertaken using a MultiSampler system capable of automatically measuring up to 50 samples at a time. 2 ml of isopropanol were used and samples measured at 50°C following a 15-minute equilibration period. Sample measurement took 60 seconds. Using Sammon mapping several samples of the two grades of isopropanol were plotted on a three-dimensional map to show a clear discrimination and close clustering of the two grades (see Figure 7.7, page 92).

AGRICULTURAL MALODOURS

There is an increasing need to assess malodours in agriculture, particularly where intensive techniques are used near to urban areas. Many of the sub-

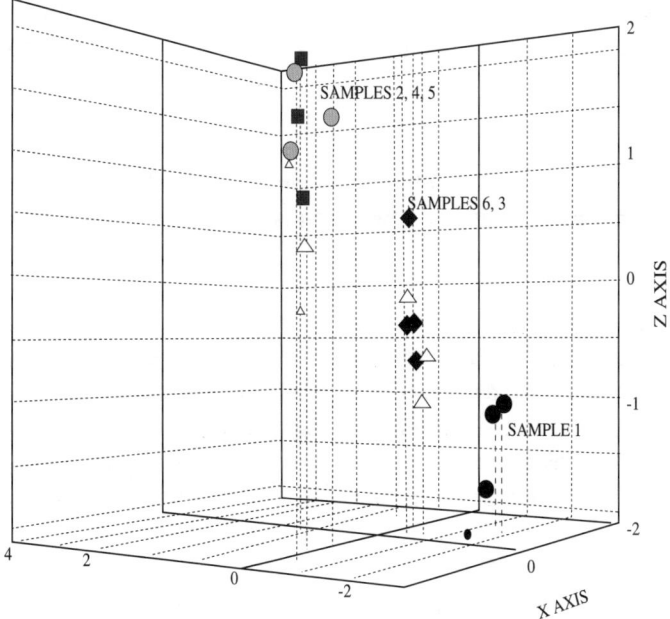

Figure 7.6   AromaMap illustrating different grades of benzyl acetate.

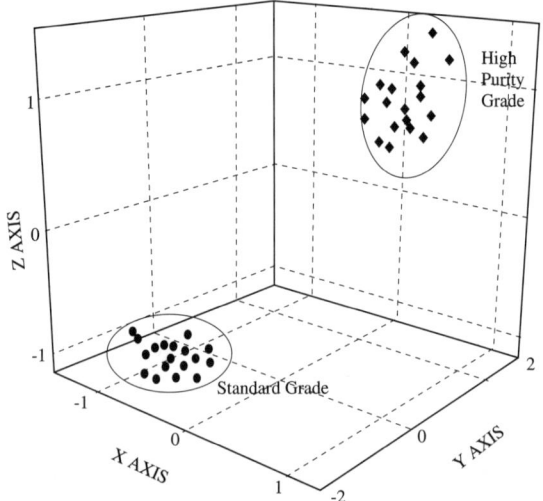

Figure 7.7   Isopropanol AromaMap.

stances produced from the anaerobic digestion of cow and pig faeces have very low human olfactory thresholds and so are perceived as odour nuisances at very low levels. These include volatile fatty acids, amines, sulphides, disulphides and mercaptans.

Traditional methods of measuring odours of this type involve the repeated dilution of the samples with clean odour-free air to 50% detection threshold by sensory panel. This is a slow, laborious and expensive system of measurement.

Experiments with pig slurry using an AromaScan system have involved comparing the main odoriferous compounds with artificially prepared pig slurry. The artificial pig slurry was made from the main constituents of slurry collected from pigs fed on controlled diets and analysed by GC/MS.

The results plotted on a two-dimensional map (Figure 7.8) show the patterns obtained for acetic acid, butyric acid, phenol, cresol, 4-ethyl-phenol, skatole, indole, ammonia and artificial pig slurry. The map shows that most of the odorous substances in pig slurry are discriminated from each other and that artificial pig slurry mapped close to indole, skatole and ammonia. The system gave reproducible responses over a four-month period and shows how it is possible to get rapid objective results for the measurement of agricultural malodours.

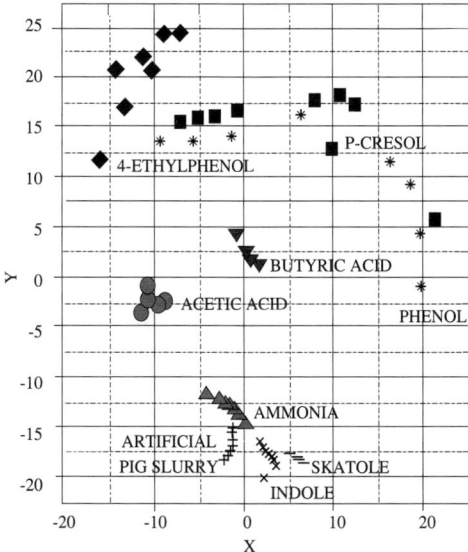

Figure 7.8   AromaMap showing the main odorous compounds in pig slurry compared with artificially prepared slurry.

CONCLUSIONS

Whilst current systems that measure odours and aromas electronically are not capable of detecting the subtle nuances identified by the human system, they are capable of performing routine operations that may be hazardous or are subject to variations commonly experienced by the human nose. There is considerable scope in the chemicals industry for the AromaScan system to provide a valuable primary screening tool so that the more complex sensory and analytical systems may be used for higher value tasks.

8. DESIGN OF DISTRIBUTED EFFLUENT TREATMENT SYSTEMS

Robin Smith and Eric Petela

INTRODUCTION

The concept of best available techniques not entailing excessive cost (BAT-NEEC) has tended to concentrate on end-of-pipe effluent treatment and on waste minimization at source. However, there is another important element to the complete water-using system that is often neglected — that of the overall strategy for effluent treatment. This chapter assumes that BATNEEC has been applied to minimization of waste generation at source and the problem is one of disposal of residual aqueous waste.

Effluent treatment in the process industries is most often carried out by collecting aqueous effluents into a common sewer system, together with utility effluents such as cooling tower blowdown, and treating them in a central on-site facility (see Figure 8.1 on page 96). This centralized facility most often uses biological treatment but might include primary and even tertiary treatment in addition to the secondary (biological) treatment. On smaller sites the effluent is often collected together, given some pretreatment, and then sent off-site to a municipal facility. The stages of treatment needed depend on the contaminants, their concentrations and the discharge regulations.

The fundamental characteristic of centralized treatment systems is that they treat large volumes of effluent at relatively low concentrations. Indeed the mixing which occurs in the sewer is usually prompted by the desire to dilute the more concentrated streams on the site to bring their concentrations down to a level suitable for biological treatment. In Figure 8.1, potentially highly contaminated streams from process operations are mixed with streams from utility plant which often either need no treatment or minimal treatment.

The problem with centralized treatment is that combining two waste streams which require different treatment technologies leads to a cost of treating the combined streams which is likely to be more expensive than individual treatment of the separate streams. This situation results from the fact that the capital cost of most waste treatment processes is highly dependent on the total flow of waste water. Also, operating cost for treatment usually increases with decreasing concentration for a given mass load of contaminant to be treated. On the

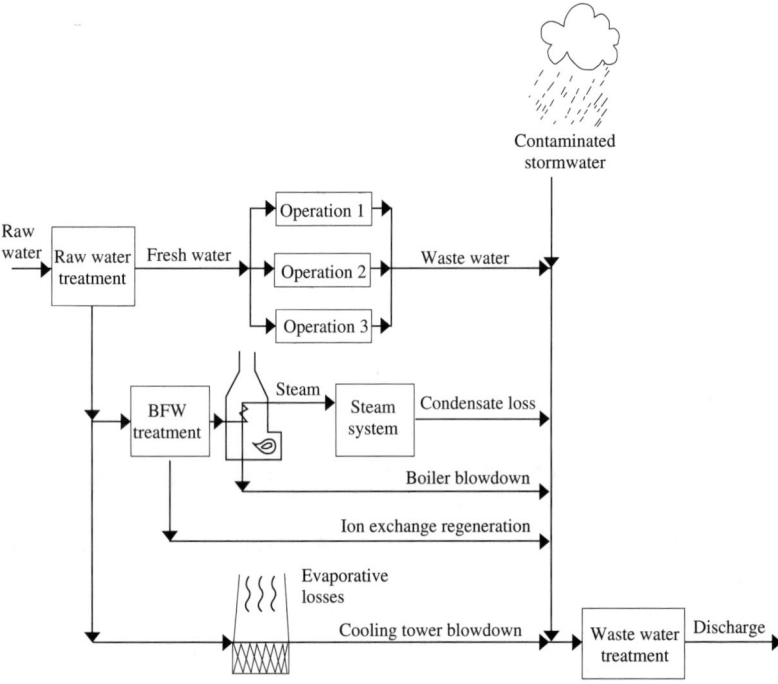

Figure 8.1   Water use and treatment on a typical site.

other hand, if two waste streams require exactly the same treatment then it makes sense to combine them to obtain economies of scale for the equipment.

A more sensible approach would be one in which effluents are combined or segregated for treatment, depending on which method is most appropriate. If this strategy is adopted then the treatment system becomes distributed rather than centralized.

The concept of distributed, or segregated, effluent treatment is not new. The basic concept has been known for many years to be a cost-effective alternative to centralized treatment. However, there has been widespread reluctance to accept the concept on the part of designers, despite the considerable potential advantages. One problem is that treatment is still overwhelmingly thought of in terms of end-of-pipe. Placing a centralized effluent treatment system on the end of the pipe is the easy, no-fuss solution. Another problem has been that until recently no design methods have been available for the systematic design of distributed effluent treatment systems. However, methods have recently been developed for design[1]. The designs which emerge raise some interesting questions for BATNEEC.

TARGETING EFFLUENT TREATMENT COSTS

The new method starts by setting targets for effluent treatment. Consider first a simple situation where a single parameter only is to be accounted for, such as chemical oxygen demand (COD). Table 8.1 presents the data for a simple problem consisting of four streams. The concentration of the final effluent must be at or below 50 ppm for discharge to the environment.

Figure 8.2(a) on page 98 shows the four effluent streams from Table 8.1 represented on concentration versus mass load axes. Whilst Figure 8.2(a) shows the problem that needs to be solved by the effluent treatment, the streams are being considered in isolation. What is needed is a view of the effluent streams as a system and not four individual streams. Figure 8.2(b) presents such a picture. The composite effluent treatment curve in Figure 8.2(b) is constructed by combining the mass loads of the individual streams in Figure 8.2(a) in the same concentration ranges. The composite curve gives an overall picture of the effluents to be treated in both concentration and mass load removal.

Figure 8.2(b) defines how much mass load needs to be removed in each concentration range in order to achieve the final effluent quality of 50 ppm. In order to achieve the effluent clean-up requirement an appropriate treatment process needs to be defined. It is possible to specify the performance of the treatment process in different ways[1]. Perhaps the simplest specification is to define

TABLE 8.1
Flow rates and inlet conditions of streams to be treated

Stream	C_{in}, ppm	Flow rate, t d^{-1}
1	300	20
2	200	60
3	100	50
4	75	40

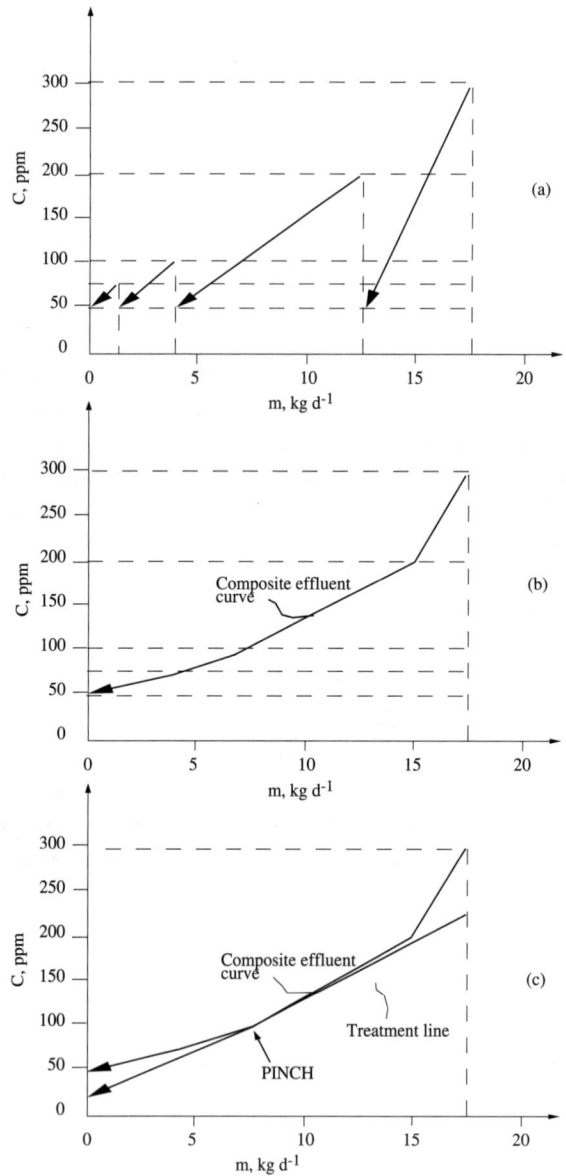

Figure 8.2   Targeting distributed effluent treatment for the simple example.
(a) Effluent stream data; (b) The stream data can be combined to produce a composite
effluent curve; (c) Matching a treatment line against the effluent composite curves sets
the target.

the outlet concentration. Alternatively, a removal ratio can be specified, defined by:

$$\text{Removal ratio} = \frac{f_{in}\, C_{in} - f_{out}\, C_{out}}{f_{in}\, C_{in}}$$

Let us also suppose that for this problem the removal ratio is 0.9. Figure 8.2(c) shows the composite effluent treatment curve with a treatment line matched against it. The treatment line shows what happens in the effluent treatment process in terms of inlet concentration, outlet concentration and the mass removed. In Figure 8.2(c) the treatment line has been adjusted to be as steep as possible. This is consistent with minimizing the flow rate through the treatment process, whilst removing the required mass load with a removal ratio of 0.9. The specification of the treatment line must obviously take account of the limitations of the treatment process under consideration in terms of maximum inlet concentration, minimum outlet concentration, removal ratio and so on. Given these limitations, the line with the maximum slope gives the minimum treatment flow rate or *target*. The point which limits the treatment line is the *pinch* for the system and is important in deciding how to design in order to achieve the target. In the case of this simple example, the target flow rate is 94.4 t d^{-1} compared with an overall flow rate of 170 t d^{-1}. This implies that only 56% of the total effluent flow needs to be treated. The following section shows what this means in practice.

The approach is readily extended to effluent problems which require multiple treatment processes and to multiple contaminants[1].

DESIGN TO ACHIEVE THE TARGETS

The construction in Figure 8.2(c) provides a target for the minimum flow rate to be treated given the environmental discharge limit and the performance of the treatment process under consideration. But how is the target achieved in practice? The method to achieve the target in design is implicit in the construction of Figure 8.2(c), in particular the role of the pinch. To achieve the target from Figure 8.2(c) the streams must be divided into three groups according to their initial concentration. The first group of streams have initial concentrations starting above pinch concentration (100 ppm in the simple example). The second group have initial concentrations equal to pinch concentration and the third group have initial concentrations starting below pinch concentration. To achieve the target from Figure 8.2(c), three design rules must be obeyed[1]:

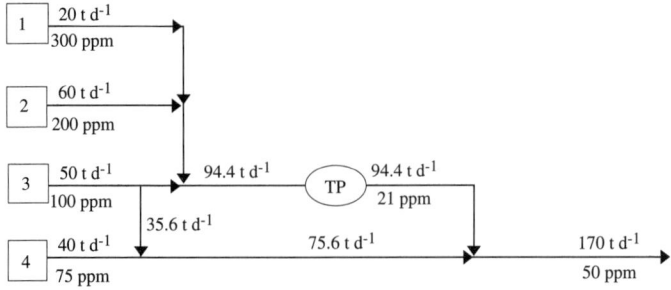

Figure 8.3   Design to meet the target for the simple example.

- effluent streams with initial concentrations above pinch concentration must pass through the treatment process;
- effluent streams with initial concentrations equal to pinch concentration may partially bypass the treatment process;
- effluent streams with initial concentrations below pinch concentration should bypass the treatment process.

The application of these rules to the simple problem leads to the design in Figure 8.3 which achieves the target set in Figure 8.2(c). Streams 1 and 2 are fully treated and Stream 3 is partially treated, whereas Stream 4 is not treated at all.

MULTIPLE CONTAMINANTS

So far this discussion has been based on the need to treat a single or 'grouped' contaminant such as COD. What happens if a number of specified contaminants need to be treated, such as individually specified organic species, suspended solids, pH, sulphate and colour? In fact the approach remains broadly the same. Each contaminant is targeted for in turn taking into account the environmental limit for that contaminant and the performance of the treatment processes relevant to it. The same treatment processes may or may not have been used to deal with the different contaminants. The target for an individual treatment process will be the maximum flow rate for that treatment process across all contaminants. Following the design rules for each contaminant leads to a design for each contaminant. The individual designs for each contaminant are then merged to produce a common design which retains all of the features of the designs for the individual contaminants.

A recent application of the technique is now reviewed briefly.

100

CASE STUDY[2]

The Monsanto Plant in Newport, Wales, recently applied the method to its effluent treatment problem. At this site, effluent from each of the seven process plants was collected together, adjusted for pH and then discharged to the River Severn Estuary. The National Rivers Authority had indicated that current discharge levels would be unacceptable in the long term. The objective was set to achieve a 90% reduction in the COD load discharged.

First the standard solution of centralized treatment was explored. For a number of reasons an oxygen-based aerobic digestion system was considered to be the most suitable. The capital cost of the new centralized treatment was estimated to be US$15 million. Unfortunately, this was considered to be an unacceptable capital expenditure. To make matters worse, the treatment process would also incur high running costs. An alternative solution was sought.

The problem was tackled from three angles using a combination of waste minimization at source[3], water reuse[4] and distributed effluent treatment[1]. In exploring options for waste minimization and water reuse the resulting effluent treatment problem for each option could be evaluated directly using the effluent targeting method already described. For each option the most appropriate treatment method could readily be chosen and sized without having to resort to repeated design.

The result was that a difficult environmental problem could be solved not with an unproductive investment of US$15 million, but with a total investment of US$3.5 million which would also bring operating cost savings of US$1 million. This impressive result was awarded a *TCE* Excellence in Safety and Environment Award in the Water Specialist Category for 1995.

DISCUSSION

Following a philosophy of distributed effluent treatment reduces the volume of effluent to be treated and reduces effluent treatment costs. Designs such as those in Figure 8.3 achieve their required final effluent concentration only after the treated and untreated effluents have been mixed together. This appears at first sight to rely on dilution as a solution, which is unacceptable. However, as far as BATNEEC is concerned this requires careful interpretation. Such an interpretation is based on the prejudice that the best way to treat effluents is to mix them together for treatment. Centralized effluent treatment also uses mixing, but before the effluents have been treated. Mixing in this way for centralized treatment reduces the potential for material recovery through the treatment system. Segregation of the effluents for treatment encourages the application of treatment processes which are different from those that would be used if the effluents had simply been mixed and treated centrally. Treating the effluents whilst they are still concentrated (and before mixing the contaminants) in a distributed treatment system often leads to the application of processes which allow recovery of useful materials. This proved to be an important feature of the case study.

Distributed effluent treatment opens up options which are closed down as soon as effluents have been mixed together. By segregation of the effluents the designer can reduce costs and increase the potential for material recovery. Distributed (or segregated) effluent treatment should therefore be viewed as BATNEEC just as much as waste minimization at source and the most appropriate choice of treatment technique. It is therefore argued that BATNEEC should entail a three-stage philosophy. The first stage is waste minimization at source, followed by determination of effluent treatment strategy and finally choice of the most appropriate treatment processes.

NOTATION

C concentration
C_{in} inlet concentration
C_{out} oulet concentration
f_{in} inlet flow rate
f_{out} outlet flow rate
m mass load of contaminant

REFERENCES IN CHAPTER 8

1. Wang, Y-P. and Smith, R., 1994, *Chem Eng Sci*, 49: 3127–3145.
2. Smith, R., Petela, E. and Howells, J., 1996, *The Chemical Engineer*, No 606, 21–23.
3. Smith, R. and Petela, E., 1991, *The Chemical Engineer*, No 506, 24–25.
4. Wang, Y-P. and Smith, R., 1994, *Chem Eng Sci*, 49: 981–1006.

9. NOₓ — NO PROBLEM: PROCESS IMPROVEMENT AND POLLUTION PREVENTION WITH HYDROGEN PEROXIDE

Stephen Woods, Phillip Wyborn, Gwenda McIntyre and Sarah Colgan

Emissions of oxides of nitrogen (NO and NO_2), or NO_x, cause serious pollution problems — contributing to acid rain and to ground level ozone pollution. Whilst only a small proportion of NO_x releases to the environment arise from industrial processes, emissions from industrial sources cause local pollution problems and are targeted for reduction. Strict consent limits for NO_x emissions from various industrial processes are being applied across Europe.

Technologies based on hydrogen peroxide are increasingly being recognized as offering cost-effective, efficient and environmentally favourable solutions to NO_x emission problems from industrial sources. These technologies are utilized both to prevent NO_x emissions and to improve process efficiency — for example, ex-process acid recovery, in-process acid recovery and nitric acid replacement.

This chapter covers the science and technology of each application and details typical operating conditions and process improvements which may be achieved.

INTRODUCTION

NOₓ — THE PROBLEM

The emissions of acid-forming oxides of nitrogen — known collectively as NO_x — cause serious pollution problems[1]. NO_x emissions contribute to acid rain and ground level ozone pollution which yields the photochemical smogs now occurring regularly during the summer in many European cities. Aside from damage to plants and buildings, various components of NO_x are poisonous to humans and NO_x emissions may be implicated in respiratory illnesses such as asthma.

The acid-forming oxides of nitrogen include nitric oxide (NO), nitrogen dioxide (NO_2), dinitrogen tetroxide (N_2O_4) and dinitrogen trioxide (N_2O_3). Dinitrogen oxide (N_2O), whilst being an oxide of nitrogen, is not acid-forming and is very stable in the environment. N_2O is therefore generally not included as a NO_x emission when setting or measuring emission limits for industrial facilities.

The vast majority of NO_x released to the environment arises from combustion and transport processes — coal-fired power stations, municipal incinerators, motor vehicle exhaust emissions and so on. Less than 10% of all NO_x releases are from industrial sources — the majority from nitric acid producers and users, and from the processing of minerals and metals at high temperature. Despite the relatively small contribution of industrial processes to the NO_x release inventory, NO_x emissions from industrial processes have been targeted for reduction across the European Union and strict consent limits are now being applied in many member states (see Table 9.1).

TABLE 9.1
Industrial processes NO_x emission limits

Country	Notes	NO_x limit expressed as NO_2	
		mg Nm^{-3}	ppm
UK	IPC regulated processes		146
Italy	Italian legislation lists 54 different industrial processes and their corresponding NO_x limits	500–1000	270–530
Germany	TA LUFT legislation	500–1000	270–530
Portugal	Other limits set to specific industries	1500	800
Belgium	Applies to all industrial emissions	500	270
Ireland	Air quality standard to be attained over the period of a year	200	110
Greece	Nitric acid production plants:		
	— new plants;	5*	
	— old plants	8*	
Spain	H_2SO_4 production:		
	— new plant	1000	530
	— old plant	3000	1600
	HNO_3 production:		
	— new plant	410	220
	— old plant	3200	1710
	Other industrial processes		300

* mg NO_2 per tonne nitric acid produced

Many industrial processes produce a mixture of NO_x species — NO and NO_2 are the predominant species. The relative concentrations of NO and NO_2 are process dependent. Thermal or high temperature processes tend to produce a higher level of NO. Hence NO is known as 'thermal NO_x'.

Whilst NO is colourless, NO_2 has a characteristic red/brown colour which is clearly visible in air at concentrations of 250 ppmv and above — allowing unabated NO_x emissions with a significant NO_2 component to be identified easily.

As indicated already, industrial NO_x emissions arise from many different processes. Specific examples are:

- manufacture of nitric acid. This involves the catalytic oxidation of ammonia to NO, subsequent aerial oxidation of NO to NO_2 and absorption of gaseous NO_2 in water to form the acid. Significant product loss occurs via NO_x emissions from the final absorption stage;

- use of nitric acid in metal, metal alloy and mineral processing. This involves the use of nitric acid to remove scale and to dissolve surface metal oxide layers to brighten or polish the metal or to dissolve the metal, alloy or mineral itself. The common chemical reaction involved in all these processes is oxidation of the metal and reduction of nitric acid to NO and NO_2;

- use of nitric acid in organic and inorganic chemicals production. This involves the use of nitric acid as an oxidant (for example, in preparation of ferric iron salts from ferrous iron) or as a nitration agent (for example, in the preparation of aminophenols). Reduction of nitric acid during such processes yields an off-gas rich in NO and NO_2.

NO_x — THE SOLUTION

Hydrogen peroxide (H_2O_2) based technologies are increasingly being recognized as offering cost-effective, efficient and environmentally favourable solutions to NO_x emission problems from industrial sources[2,3]. Hydrogen peroxide based technologies are utilized both to prevent NO_x emissions and to improve process efficiency. For example:

- nitric acid recovery — applied as a component of the scrubbing liquor in the wet scrubbing of gaseous emissions, hydrogen peroxide allows recovery of nitric acid, can eliminate NO_x emissions and can reduce process acid requirements;

- in-process acid recovery — the incorporation of hydrogen peroxide within the process liquors, to act as an oxidant for *in situ* regeneration of nitric acid, reduces acid requirements and can improve the overall process;

• nitric acid replacement — use of hydrogen peroxide in place of nitric acid as an oxidant is an alternative for some applications. Nitric acid free processing is a commercial alternative in the copper and steel industry.

HMIP has recognized the benefits of hydrogen peroxide for reducing NO_x emissions when applied to wet scrubbers and within metal pickling baths (see Appendix 9.1 on page 113).

HYDROGEN PEROXIDE — THE PRODUCT

The search for effective, clean and economic oxidation methods is a challenge facing chemists throughout the world. Hydrogen peroxide and its derivatives are versatile oxidants, well placed to convert a variety of compounds into useful products.

There are several reasons why hydrogen peroxide and its derivatives have grown in importance in conditions where oxidation is required. They include the following:

• efficient, high volume production processes capable of producing hydrogen peroxide at various concentrations and at high purities are available in many parts of the world;

• hydrogen peroxide has a high active oxygen content, so on a weight-for-weight basis it is more efficient than other oxidants such as potassium dichromate or permanganate;

• hydrogen peroxide is stable in storage; when handled correctly, it loses less than 1% of its active oxygen content per year;

• hydrogen peroxide oxidations can proceed under relatively mild conditions and give rise to specific products;

• when the oxidizing power of hydrogen peroxide is spent, water and oxygen are the only by-products.

Transition metals catalyse the decomposition of hydrogen peroxide to water and oxygen, and it is preferable to operate at temperatures below 70°C to minimize this effect.

There are several features of current industrial oxidation processes where the use of hydrogen peroxide offers environmental advantages over alternative oxidants. In order of priority these are seen as:

• replacement of stoichiometric metal oxidants (recovery, disposal, environmental persistence problems);

• replacement of halogens (effluent, environmental persistence problems);

• avoidance of salt by-products (some effluent problems).

At present, industry mainly uses processes which originated before effluent considerations were an integral part of process design. 'End-of-pipe' treatments are therefore being used in the short term to enable existing processes to operate under ever more stringent environmental legislation. Hydrogen peroxide can be utilized both in 'end-of-pipe' and 'in-process' treatments — allowing the user to choose the most appropriate route for pollution control and prevention according to site constraints and requirements. Technical, financial and safety aspects of any proposed treatment must be considered in deciding upon the most appropriate technology for a particular process.

PEROXYGEN TECHNOLOGIES FOR NO$_x$ CONTROL

Hydrogen peroxide can be applied in several ways to control NO$_x$. The technologies involved are:

- nitric acid replacement;
- gas scrubbing;
- NO$_x$ suppression.

The chemistry[4] involved in the use of hydrogen peroxide for gas scrubbing NO$_x$ and suppression of NO$_x$ in process is essentially the same and is outlined in Appendix 9.2 on page 114. In *gas scrubbing*, the reactions follow the absorption of NO$_x$ gases into the aqueous phase. The hydrogen peroxide, dissolved in nitric acid, oxidizes the NO$_x$ to nitric acid. Typically *for NO$_x$ suppression*, hydrogen peroxide rapidly oxidizes the nitrous acid (HNO$_2$) formed by the reduction of the nitric acid by a metal, before it has a chance to decompose to NO$_2$, NO and H$_2$O[5,6]. In both of these applications hydrogen peroxide has the advantage that its reaction with NO$_x$ produces nitric acid which can invariably be recycled back into the process as a reagent.

As with all industrial processes, variants on processes for NO$_x$ control are frequently subject to patent protection, and measures may need to be taken to avoid infringement.

The technology of each of the three processes is described in detail below, together with details of typical operating conditions and process improvements which may be achieved.

NITRIC ACID FREE PROCESSING

In the manufacture of stainless steel components, after the metal has been mechanically worked, the internal stresses induced by the working must be relieved by annealing the material in a furnace at high temperature. This annealing process builds up a series of layers on the surface of the stainless steel. The removal of these layers is achieved by mechanical and chemical means. The chemical process, known as 'pickling', is traditionally carried out using a mixed acid solution containing hydrofluoric and nitric acid. Pickling removes the chromium depleted layer at the surface of the stainless steel oxidatively, with complimentary reduction of the nitric acid to NO and NO_2.

The use of the nitric acid serves two functions, as both a mineral acid and an oxidant. As hydrogen peroxide is also an oxidant, it can be used in the pickling process in conjunction with an alternative mineral acid.

Solvay Interox has developed a pickling process based on an $HF/Fe/H_2SO_4/H_2O_2$ system[7]. The hydrogen peroxide has to contain stabilizers and other additives to reduce decomposition due to contact with the metal and carry out the pickling. Solvay Interox developed a stabilized hydrogen peroxide formulation OXYFIRST Steel for this application. This, in combination with the other components of the pickling liquor and Solvay Interox's process recommendations, provides an efficient nitric acid free pickling process.

Adoption of the OXYFIRST Steel process for stainless steel pickling allows considerable process improvements and cost reduction over a traditional HF/HNO_3 process. For example, based on the traditional HF/HNO_3 system cost of 1.00:

Initial chemical costs:
HF/HNO_3 1.00
$HF/Fe/H_2SO_4/H_2O_2$ 1.47

Pickling bath life:
HF/HNO_3 1.00
$HF/Fe/H_2SO_4/H_2O_2$ 4.00

Bath make-up times:
HF/HNO_3 1.00
$HF/Fe/H_2SO_4/H_2O_2$ 0.12

Effluent disposal costs:
HF/HNO_3 1.00
$HF/Fe/H_2SO_4/H_2O_2$ 0.12

In this case the fourfold increase in pickling bath lifetime, associated reduction in time required for preparation of new baths and reduced effluent disposal cost makes OXYFIRST Steel technology an attractive alternative to a traditional system, despite the higher initial chemical costs.

Nitric acid free processing can also be achieved in processes where metals are dissolved in acid mixtures containing nitric acid in which the nitric is present only as the oxidant. In such a system the oxidizing power of the nitric acid can be replaced with hydrogen peroxide and an alternative acid used to provide the anion.

In chemical synthesis nitric acid can be used as an acid, oxidant and/or a source of nitrogen. Clearly, when it is used as a nitrogen source for a nitration product or intermediate, hydrogen peroxide is not a viable alternative. However, if the process is analogous to the stainless steel process, where the nitric acid is being used as both an oxidant and acid, then the combination of hydrogen peroxide and a suitable alternative acid is a potential synthesis route.

GAS SCRUBBING

Gas scrubbing, in general, involves the removal of hazardous or toxic gases from a gas stream. The pollutant is removed from the gas and collected in a liquid phase. Gas is collected from a process, and sent to a scrubber where the gas passes up through a liquid that is being run down the scrubber. Packing provides surface for the contact of the gas with the liquid. Simple packings are cheaper than the intricate packings, but the more intricately designed materials tend to give better scrubbing efficiency. Figure 9.1 on page 110 shows a schematic of a packed tower scrubber.

In processes using nitric acid, NO_x is emitted from the process, the gas is cooled and passed through a countercurrent stream of 5–30% HNO_3 (usually 20–30%) and (normally) < 5% H_2O_2. Removal of NO_x is usually 80% which is considered very efficient. The actual efficiency of the system, however, is dependent on the $NO:NO_2$ ratio (because NO_2 is more readily absorbed into the aqueous phase and eliminated), and parameters such as gas flow. Nitric acid is used in conjunction with hydrogen peroxide because nitric acid helps oxidize the more stubborn NO in both the gaseous and the liquid phases to NO_2[8,9].

Additions of hydrogen peroxide to the scrubber liquor can be made either continuously or in batches. The nitric acid produced — which normally contains low levels of hydrogen peroxide — may safely be reused in most cases. This is advantageous in, for example, metal pickling processes where the

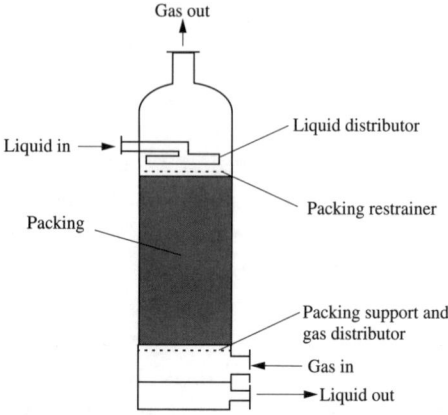

Figure 9.1   Schematic of a gas scrubbing system.

peroxide in the acid suppresses NO_x emissions *in situ*. Alternatively, the scrubbing system can be designed to produce peroxide free acid.

Stoichiometrically, the amount of hydrogen peroxide required to treat NO_x is 0.37 kg of H_2O_2/kg of NO_2 and 1.7 kg of H_2O_2/kg of NO. These quantities are the theoretical amounts and in practice there is a slight excess of hydrogen peroxide required due to decomposition.

Solvay Interox has developed a gas scrubbing simulation computer program which takes parameters that affect gas scrubbing into consideration and gives a suggested design for a scrubber using hydrogen peroxide (height, diameter, etc). The system can also be used to input the parameters of an existing plant to determine the expected NO_x outlet values if it were converted to use a nitric acid/peroxide scrubber liquor. By selection of the correct acid/peroxide strengths it can be determined if the existing plant can meet the required discharge limits, or what modifications or additions are required. A gas scrubbing questionnaire issued to potential customers assists in obtaining the data for the computer program.

NO$_x$ SUPPRESSION
This technology uses hydrogen peroxide to oxidize the nitrous acid (HNO_2) before it decomposes to NO_x. Under normal operating conditions the reaction of hydrogen peroxide with HNO_2 is marginally faster than the rate of hydrogen peroxide decomposition. To ensure that the oxidation reaction occurs preferentially, very good mixing of the hydrogen peroxide into the process liquor is essential.

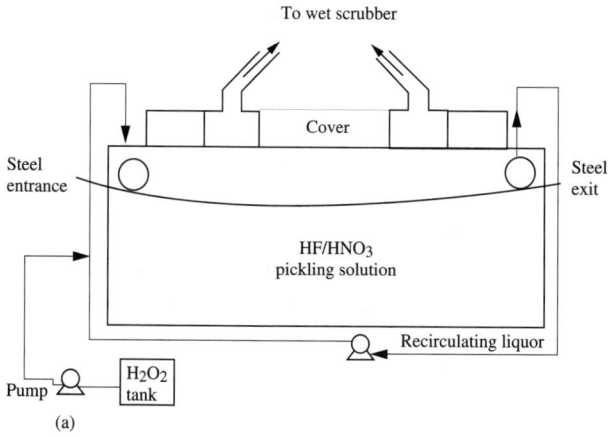

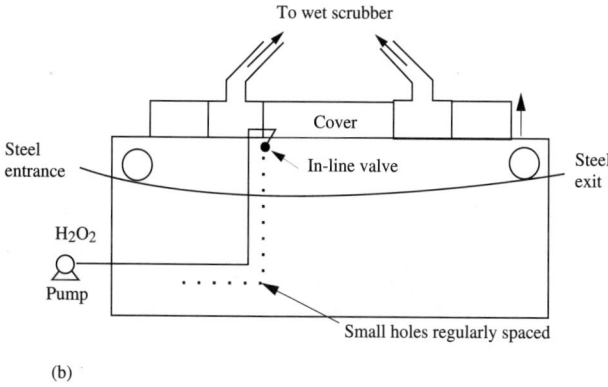

Figure 9.2 Schematic of a suppression system: (a) with H_2O_2 addition via recirculation loop; (b) with H_2O_2 addition directly to the pickling bath.

Hydrogen peroxide can be introduced into a bath by injection into a recirculation loop (see Figure 9.2(a)). The contents of the bath should ideally be recirculated at a rate of about five bath changes per hour. If no recirculation loop is available then a simple hydrogen peroxide dosing arrangement can be used.

This features a pipe drilled with small holes at regular intervals strategically positioned in the bath (see Figure 9.2(b)).

In stainless steel manufacturing, the steel is normally pickled in a mixture of HF and HNO_3. When hydrogen peroxide is used for NO_x suppression, up to 90% NO_x is typically removed (sometimes up to 95% suppression) and the quality of the stainless steel is not adversely affected. Regeneration of nitric acid can lower nitric acid consumption by 20–30% and significantly reduce the losses of HF via entrainment in gaseous NO_x emissions.

Stainless steel grades vary based upon the amount of carbon, iron and chromium (and other metals) they contain. Series 300 steels (austenitic) are generally endothermic in pickling character and 400 series steel is generally exothermic. The type of steel has a great impact on the amount of NO_x liberated. This has been found at a number of plants where compliance to local regulations has necessitated the commercial operation of a NO_x suppression process based on hydrogen peroxide.

As every process is different in terms of the type of steel or metal being processed and in terms of the bath conditions, the amount of peroxide to be dosed varies. Optimal peroxide dose requirements are usually determined via full-scale trials on the process.

CONCLUSIONS

Hydrogen peroxide based technologies offer cost-effective, efficient and environmentally favourable solutions to NO_x emission problems through prevention (via nitric acid replacement or *in situ* NO_x suppression) or through abatement (via scrubbing of process off-gases). All three of these hydrogen peroxide based technologies have been proven in industrial trials and are in commercial use.

REFERENCES IN CHAPTER 9
1. Slater, D., 1995, Air dye, *The Chemical Engineer*, 29 June 1995.
2. *ENDS Report 235*, August 1994, 32.
3. *ENDS Report 208*, May 1992, 12–13.
4. Deo, P.V., 1988, The use of hydrogen peroxide for the control of air pollution, *Chem Prot Environ*, 34: 275–292.
5. Buck, M., Clucas, J., McDonogh, C. and Woods, S., 1991, NO_x removal in the stainless steel pickling industry, *US Chem Oxid Proc Int Symp*, 78–88.
6. Karlsson, H.T., Nilsson, L-I.O., Schutsky, G. and Bjerle, I., 1984, Control of NO_x in steel pickling, *Environmental Progress*, 3 (1): 40–43.
7. International Patent Application *WO 93/08317*.

8. Thomas, D., Brohez, S. and Vanderschuren, J., 1996, Absorption of dilute NO_x into nitric acid solutions containing hydrogen peroxide, *Trans IChemE*, 74 (B1): 52–58, and references therein.
9. Thomas, D. and Vanderschuren, J., 1993, Flue gas treatment — absorption of nitrogen oxides in solutions containing oxidising agents, *Recénts Progrès en Génie des Procédés*, 7 (30): 369–374 (Etudes et Conception d'Equipements).

APPENDIX 9.1 — HMIP RECOGNITION OF THE BENEFITS OF HYDROGEN PEROXIDE FOR NO_x CONTROL

Of gaseous abatement techniques for control of oxides of nitrogen from organic nitration processes:

'Improved scrubbing performance, in some circumstances, can be expected by using solutions of ammonium hydroxide, urea or hydrogen peroxide; the latter offering the best performance and minimal effluent disposal problems.'
Chief Inspector's Guidance to Inspectors, Note IPR 4/12 (page 39)

Of metal pickling chemistry:

'A new development in stainless steel pickling is the dosing of the acid solution with hydrogen peroxide. This reduces the nitrogen oxides emissions and the quantity of nitric acid used.'
Chief Inspector's Guidance to Inspectors, Note IPR 4/11 (page 29)

Of metal pickling chemistry:

'Hydrogen peroxide treatment of the pickling acid normally gives a 20 to 25% saving in nitric acid consumption.'
Chief Inspector's Guidance to Inspectors, Note IPR 4/11 (page 30)

Of abatement systems for metal pickling:

'Addition of hydrogen peroxide to the bath can reduce NO_x emissions by oxidising the nitrogen monoxide formed to nitrogen dioxide which is re-absorbed within the bath acid. Emissions of NO_x from the bath can be reduced by up to 70% although the economics of this technique will need to be assessed in relation to the specific application.'
Chief Inspector's Guidance to Inspectors, Note IPR 4/11 (page 46)

Of abatement techniques for dissolution of metals:

'Wet scrubbing systems based on packed towers or high efficiency venturi may be employed. The use of reagents such as hydrogen peroxide in the circulating solution will, most likely, be necessary to achieve the required NO_x emission limits.'
Chief Inspector's Guidance to Inspectors, Note IPR 4/11 (page 48)

APPENDIX 9.2 — THE CHEMISTRY OF NO_x ABATEMENT AND SUPPRESSION WITH HYDROGEN PEROXIDE

NO	=	nitric oxide
NO_2	=	nitrogen dioxide
HNO_2	=	nitrous acid
HNO_3	=	nitric acid
N_2O_3	=	dinitrogen trioxide
N_2O_4	=	dinitrogen tetroxide

NO

$NO(g)$	$\rightarrow$	$NO(aq)$
$NO(aq) + H_2O_2$	$\rightarrow$	$NO_2(aq) + H_2O$
$3NO_2 + H_2O$	$\rightarrow$	$2HNO_3 + NO$
$2NO + 3H_2O_2$	$\rightarrow$	$2HNO_3 + 2H_2O$

NO_2

$2NO_2(g)$	$\leftrightarrow$	$N_2O_4(g)$
$N_2O_4(g)$	$\rightarrow$	$N_2O_4(aq)$
$N_2O_4(aq) + H_2O$	$\rightarrow$	$HNO_2 + HNO_3$
$HNO_2 + H_2O_2$	$\rightarrow$	$HNO_3 + H_2O$
$2NO_2 + H_2O_2$	$\rightarrow$	$2HNO_3$

NO/NO_2

$NO(g) + NO_2(g)$	$\leftrightarrow$	$N_2O_3(g)$
$N_2O_3(g)$	$\rightarrow$	$N_2O_3(aq)$
$N_2O_3(aq) + H_2O$	$\rightarrow$	$2HNO_2$
$2HNO_2 + 2H_2O_2$	$\rightarrow$	$2HNO_3 + 2H_2O$

10. SUPER-STABLE NANOFILTRATION MEMBRANES OPEN THE WAY TO NEW APPLICATIONS*

Jorge Yacubowicz and Peter Crocker

INTRODUCTION

The common membrane separation processes are microfiltration (MF), ultrafiltration (UF), and reverse osmosis (RO). Nanofiltration (NF) is a pressure-driven membrane process in which low molecular weight solutes (< 1000 dalton) are retained but salts are partially or completely passed to the filtrate. This provides a range of selectivities between ultrafiltration and reverse osmosis enabling simultaneous concentration and desalting of organic solutes. The NF membrane retains solutes that UF membranes would pass, while passing salts that the RO membranes would retain.

By allowing a relatively free passage of monovalent ions, nanofiltration membranes are able to reduce the build-up of the osmotic pressure gradient contributed by the monovalent salts. As a result, high product recoveries are possible.

The separation capabilities of the different membrane processes can be characterized on the basis of the molecular weight or physical size of the substance or molecule to be separated. Figure 10.1 on page 116 shows the behaviour of water, monovalent salts, sucrose, proteins and suspended matter when applying the membrane processes discussed.

Until recently, the high selectivity of nanofiltration has been limited to aqueous applications within the pH range 2–11. However, many industrial applications require acid/base and solvent stability. This chapter describes the technical aspects of acid, base and solvent stable membranes, and some potential applications.

MEMBRANE CHARACTERISTICS

Membrane Products Kiryat Weizmann (MPW) has developed and manufactures a unique class of nanofiltration membranes (SelRO) which are characterized by

---

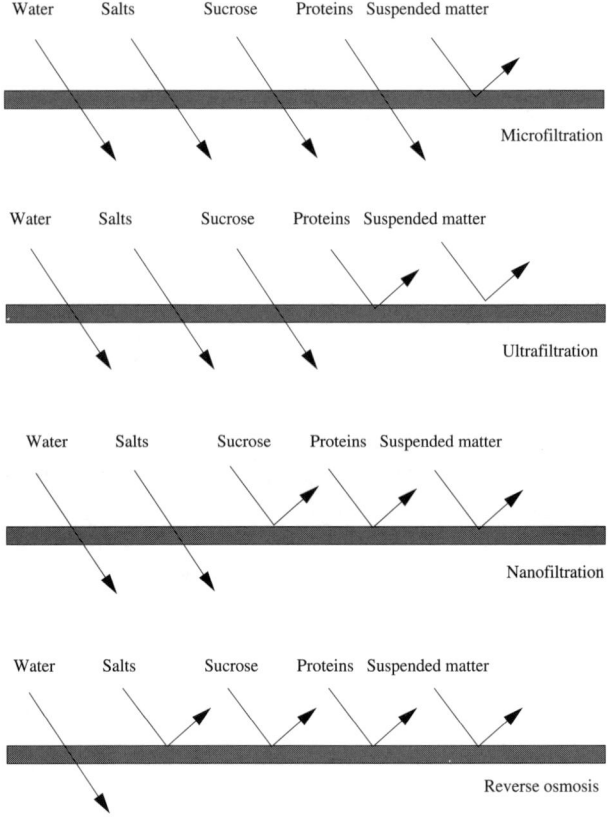

Figure 10.1 Membrane process characteristics.

an outstanding chemical and thermal stability. Table 10.1 summarizes the stability and separation characteristics of SelRO membranes. Table 10.2 shows a more detailed stability for the code 34 membrane.

The separation concept presented in Figure 10.1 can be extended by replacing the filterable salt with bases, such as NaOH, or with acid compounds, such as H_2SO_4 or HCl. Furthermore, water can be replaced with water-solvent mixtures or pure solvents. These extended separation capabilities are illustrated in Figure 10.2 on page 118. Based on the figure, the following fields of application can be identified:

TABLE 10.1
SelRO nanofiltration membranes

Code	Product name	Approximate MW cut-off	pH range	Solvent stability	Maximum temperature, °C
10	MPT–10	200	2–11		60
20	MPT–20	450	2–10	Partial	50
	MPS–21	400	2–10	Partial	45
30	MPT–30	400	0–12		70
	MPT–31	400	0–14		70
	MPT–34	200	0–14		70
	MPT–36	1000	1–13		70
	MPS–31	450	0–14		70
	MPS–34	300	0–14		70
	MPS–36	1000	1–13		70
40	MPS–44	250	2–10	Excellent	40
50	MPS–50	700	4–10	Excellent	40
60	MPS–60	400	2–10	Excellent	40

TABLE 10.2
Stability characteristics of code 34 membrane

Acid/base	Concentration	1000 hours stability
HCl	5%	Stable
	37%	Stable
H_2SO_4	15%	Stable
H_3PO_4	20%	Stable
Acetic acid	15%	Stable
HNO_3	5%	Stable
NaOH	3%	Stable
KOH	3%	Stable

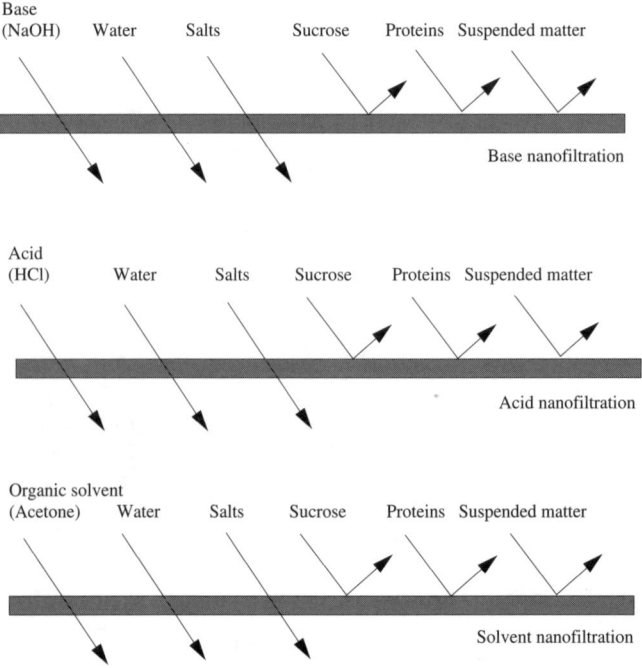

Figure 10.2   Nanofiltration process characteristics.

BASIC/ACID STREAMS

SelRO membranes can be operated within the pH range 0–14 and at temperatures up to 70°C.

Typical applications for this case are:

- recovery of spent alkaline (NaOH, KOH) streams;
- decolourization of spent mineral acids;
- removal of heavy metals from acid and basic solutions;
- concentration and demineralization of dyes in acids;
- recovery of spent electroless solutions;
- concentration/desalination of wastes from dye manufacturing;
- reduction of COD in pulp and paper waste effluents.

The case studies on pages 119–124 give further details.

ORGANIC LIQUIDS STREAMS

SelRO membranes can be operated in a wide spectrum of pure solvents and solvent mixtures, such as acetonitrile, methylene chloride, methanol, ethanol, butyl acetate and xylene.

Typical applications for this case are:
- recovery of antibiotics and peptides in organic solvents and their water mixtures;
- recovery of dissolved chemicals in the presence of organic solvents;
- purification of solvent extraction fluids to enable recycling;
- recovery of binder and pigment from paint sludge;
- removal of dissolved polymers from solvents prior to distillation;
- recovery of dissolved catalyst from organic fluids;
- recycling of hydrocarbons in cleaning processes.

The case studies on pages 119–124 give further details.

AQUEOUS STREAMS REQUIRING AGGRESSIVE CHEMICAL CLEANING

Standard nanofiltration membranes have limited chemical stability and cannot be used in certain aqueous applications due to fouling and blockage of the membrane by components in the feed.

Likewise, RO membranes frequently require extensive and costly pretreatment of the feed stream to avoid fouling.

In order to remove the blocking materials, aggressive cleaning chemicals should be used. Standard RO and NF membranes cannot withstand the required cleaning conditions. As a result, the use of RO and NF membranes has been limited to 'easy-to-clean' applications. With the introduction of ultra-stable SelRO membranes, aggressive cleaning is made possible.

Aggressive cleaning can be performed within the pH range 0–14 and at temperatures up to 70°C. Typical applications for this case are:
- the removal of dissolved silica from water;
- nanofiltration pretreatment of RO membranes;
- humic acid or COD removal from aqueous streams.

In all these cases, blocking compounds can be easily removed by a short cleaning at very low or very high pH conditions.

CASE STUDIES

COPPER RECOVERY IN ELECTROLESS SOLUTIONS BY MEMBRANE TECHNOLOGY

Electroless plating is the controlled autocatalytic deposition of a continuous film by the interaction of a metal salt (for example, copper) and a chemical reducing agent in solution. Electroless solutions contain a metal salt, a reducing agent, a pH adjuster or buffer, a complexing agent, and one or more additives to control stability, film properties and deposition rates. Reuse of electroless solutions requires filtering to remove particulate and nucleation centres that can cause bath instability and can modify the coated surface properties.

High increases in metal prices and stricter environmental regulations have driven the electroless industry to search for a technology that retrieves the unreacted complexed copper for replenishment purposes. For some years now, membrane technology has been successfully applied to achieve the tight requirements for metal-complex purification and concentration.

Electroless copper baths in commercial use are formaldehyde-based. Since the required reduction potential of formaldehyde increases with alkalinity, baths are usually operated at a pH above 12. Due to this, it is necessary to use complexing agents, such as EDTA, to prevent precipitation of cupric hydroxide.

These facts put strong requirements on the selection of the proper membrane for use in the recovery process. First, a tight membrane that has high rejection to the copper-EDTA complex should be selected. Second, and of far greater importance, the membrane should have high and long-term pH stability. Today, few membranes in the market meet this dual requirement (although the SelRO MPS–31 and MPS–34 spiral wound nanofiltration membranes do).

The recovery process for Cu–EDTA dissolved in the solution is as follows (see Figure 10.3). The membrane splits the inlet stream, containing Cu–EDTA, into two different streams. The first stream contains the majority of the Cu–EDTA complex and another, impoverished, stream which can be purified further by a second stage treatment. The second (discharge) stream contains a very low concentration (below 5 ppm) of Cu–EDTA.

RECLAMATION OF CAUSTIC FROM DAIRY EVAPORATOR AND CLEANING IN PLACE STREAMS

Cleaning in place (CIP) of production equipment such as pipelines and evaporators in dairies is routinely performed at the conclusion of each production cycle to maintain high standards of hygiene and to ensure proper functioning of the equipment. The cleaning operation involves several steps:

- a first rinse with tap water to remove suspended material;
- a second rinse with hot water to remove material that is loosely attached to the surface;
- circulation of a hot-alkaline cleaner to remove organic deposits;

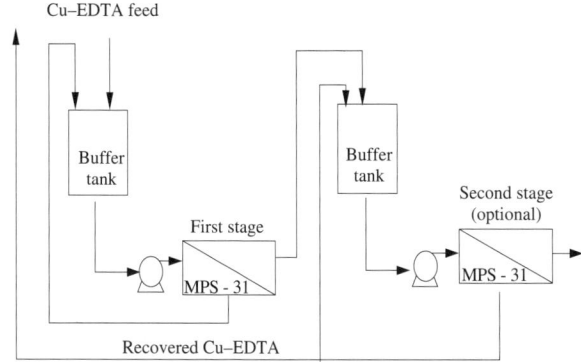

Figure 10.3   Schematic process of CU–EDTA recovery.

• an additional rinse with water.

In some CIP schemes the alkaline step is, less frequently, followed by cleaning with an acid solution:

• an acid rinse to dissolve mineral deposits;
• a water rinse and sanitation.

Since acid cleaning is performed every few days, the caustic cleaning chemicals account for a major part of the cleaning costs. The consumption of caustic can be significantly reduced by using nanofiltration membranes, as described below.

Typically, strong caustic cleaning solutions contain 2–4% NaOH or KOH and additional compounds such as antifoams and chelating agents. The alkaline cleaner removes caramelized organics, precipitated proteins, pectins and fats from the surfaces of tanks, pipes, heat exchangers and evaporators. After the cleaning step, the contaminated caustic cleaner still contains a high concentration of active caustic.

There are two major CIP concepts:

• the solution is reused 7–10 times and then the CIP tank is completely discharged;
• approximately 10% of the caustic solution is discharged daily, while the remaining volume is recycled to the CIP tank.

In both cases, the discharged caustic cannot be reused without further treatment due to its heavy load of dispersed and soluble organic contaminants. At this point, the stream must be neutralized and discharged.

Base stable nanofiltration membranes enable the removal of suspended *and* dissolved substances allowing free passage of the caustic solution.

More than 95% of the brown-burnt coloured contaminants and COD are removed from the treated caustic, permitting continuous recycling of over 95% of high quality caustic. A scheme of MPW's caustic recovery system (AlkaSave), which functions as an add-on to the caustic tank, is given in Figure 10.4.

This summary presents results of two commercial-scale installations on whey evaporator caustic and on cheese CIP caustic in which the AlkaSave system was used to purify and recover NaOH.

Evaporators in dairies

The AlkaSave installation consists of 20 tubular, SelRO MPT–34 membranes installed in TM–1228–AS modules. The TM–1228–AS module contains 18 tubes, each 3.6 m long, and has a total area of 2.6 m². The AlkaSave unit operates at a temperature of 70°C. The volume of spent caustic treated is 60,000 litres every 20 hours. The recovered caustic consists of more than 54,000 litres at the same concentration and temperature as the feed of spent caustic solution.

While the feed of spent caustic is a dark colour, the recovered caustic liquid is clear, just like newly prepared raw caustic. The solution is effective in

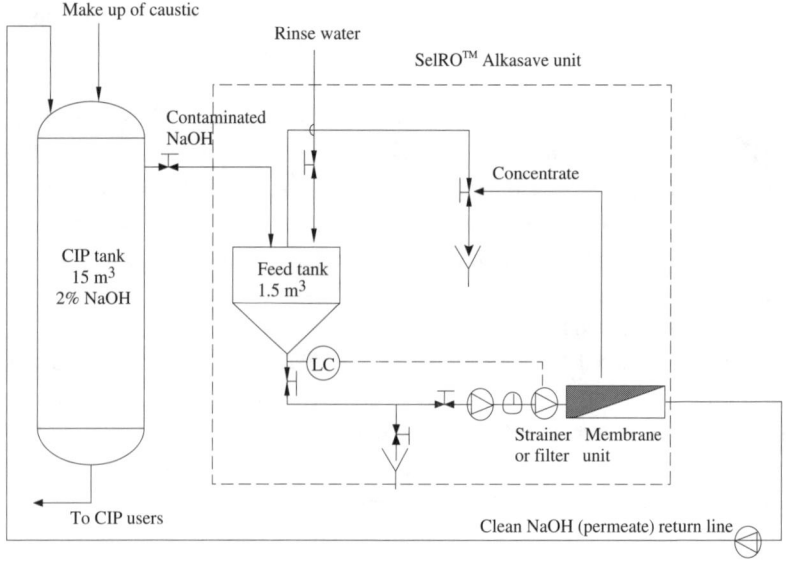

Figure 10.4   Integration of the AlkaSave process in dairies and breweries.

long-term cleaning over several months without need to dispose of it. Only about 10% make-up caustic is added as required.

PIPELINE CIP

This AlkaSave installation consists of two tubular, SelRO MPT–34 membranes installed in TM–1228–AS modules. This unit also operates at 70°C. It operates continuously over a week as an add-on to the CIP tank. Membrane rinse is done once every three days, for two hours.

The recovered caustic consists of more than 10,000 l every 24 hours. The recycled caustic has the same concentration and temperature as the spent solution feed. The solution has a clear colourless appearance and is being reused for CIP operations without need for disposal. As in the evaporator case, only a very small volume of caustic make-up is added to the CIP tank. Once washed by diafiltration to remove excess alkalinity, the concentrated organic stream can be used as an animal food additive.

The payback period is about 1.5 years. The operation does not require attendance of personnel except for washing periodically. The systems are compact in their dimensions and can be hooked as an add-on unit to CIP installations.

RECOVERY AND PURIFICATION OF ANTIBIOTICS

Antibiotics are commercially produced in a fermentation process. The broth contains approximately 4% biomass, varying concentrations of salts and about 0.1%–2% of low molecular weight antibiotics. SelRO membranes can be used in either of two ways to recover and purify the antibiotics.

The first option involves those production schemes in which organic solvent extraction equipment is used to remove antibiotic compounds from the clear filtrate of the broth. Following extraction, the dissolved antibiotics can be further concentrated by use of SelRO hydrophobic membranes that are stable to solvent. The membrane permeate — that is, the liquid going through the membrane — consists of a pure solvent which can be recycled for the next extraction batch. With this approach (an alternative to traditional distillation processes), the membrane process offers savings of up to 80% in operating costs.

In the second membrane option, hydrophilic membranes are used instead of solvent-stable ones. The aqueous membranes concentrate the dilute antibiotics in a filtered broth prior to solvent extraction. In this case, the productive capacity of the existing solvent extraction equipment is significantly increased and the required volumes of solvent are substantially reduced.

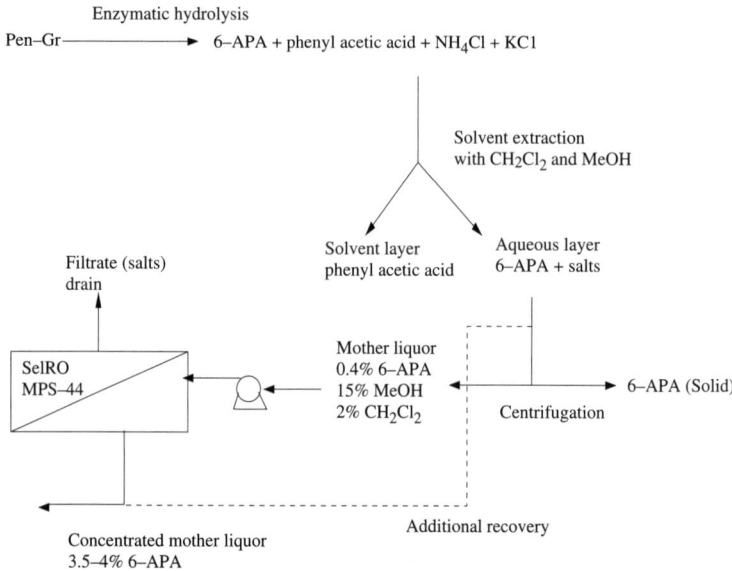

Figure 10.5   6–APA production process using membrane technology.

CONCENTRATION OF 6–APA

6–aminopeniallanic acid (6–APA) is an intermediate in the manufacturing of a synthetic penicillus with a molecular weight of 216. It is manufactured in a fermentation process or synthesized chemically.

   The typical mother liquor from a chemical reaction contains 0.4% of dissolved 6–APA, 15% MeOH and 2% methylene chloride. The membrane process concentrates the 6–APA to 5% using the solvent stable SelRO MPS–44 element (see Figure 10.5). The mass balance recovery of 6–APA is from 90–95%. The return on investment is less than one year. An alternative process is possible in which the concentration of 6–APA can reach 10% in the recovered product stream.

ADDRESS

Correspondence concerning this chapter should be addressed to Mr P. Crocker, Technical Director, Ultra Sep Limited, Langton House, 53 Langton Road, East Molesey, Surrey KT8 2HX, UK. (Direct tel/fax: +44(0)181 941 0845.)

11. SOLVENT RECOVERY
John Doyle

INTRODUCTION

Solvent recovery or recycling as a commercial service to industry is now a well-established operation in the UK. Best available techniques not entailing excessive cost (BATNEEC) can be applied to reduce the environmental impact of physical and chemical process operations, but this is only part of the cycle.

This chapter considers the complete cycle, from collection of waste to delivery of refined product. Solvent recycling is not only a technical or processing problem; numerous decisions need to be taken and problems solved to ensure that the procedures satisfy both the customers and the regulators. Solvent recycling occurs only when all the links in the chain are complete. The decision and assessment procedures are discussed, showing how an enquiry from a waste producer is treated and proposals made for the most suitable treatment. Finally, two specific examples are considered, which show how it has been possible to create a viable recycling process which removes waste from a disposal route and elevates its treatment to that of beneficial reuse. These examples demonstrate how the waste producer, the solvent recycler, the end user and the regulators must co-operate to complete the chain.

The recycling of waste industrial solvents is well established in the UK and has many similar elements to other recycling operations — such as waste paper, glass, metal, oil and so on.

Operating a solvent recovery business is like any other in that it must make a profit but — unlike most other businesses — the waste producer dictates what the raw material will be, a different end user dictates the specification of the product and the market place regularly changes the value of the recovered solvent.

All recycled solvents sell for less than the virgin product and often the recycling costs exceed the virgin price. Some products — such as organic crops or free range eggs — can be sold at a premium price, but, sadly, in the chemical industry there is little sympathy for the solvent recycler.

Very often, the factor which determines the recyclability of a waste is not the technical processing but the price of the virgin solvent or, for some less common solvents, finding an end user at all. If no-one wants the recovered product, it cannot be described as recyclable.

In mid-1994 virgin methanol cost £120–£130 per tonne. At this level the economics of recycling methanol were difficult and plentiful supplies of virgin meant little demand for reclaimed. By mid-1995 the price of virgin methanol had trebled and suddenly it was possible to recycle methanol profitably. Prices have now almost returned to their former level and once again the volume of methanol being recycled has declined. Solvent recycling may often be the best environmental option and can be performed using the best available technique, but unless it is profitable and can find a customer, it will not take place.

Before specific examples are considered, it is necessary to understand the options for recycling and the cost involved.

THE HIERARCHY OF WASTE MANAGEMENT

Figure 11.1 shows a simplified version of the accepted decision-making process for managing waste. Usually, this hierarchy also coincides with a cost-effective solution hierarchy. Sometimes the availability of a lower cost landfill option may discourage the reuse or recycle option. The forthcoming landfill tax will encourage the recycling options.

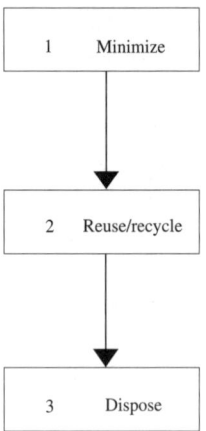

Figure 11.1 A simplified hierarchy of waste management.

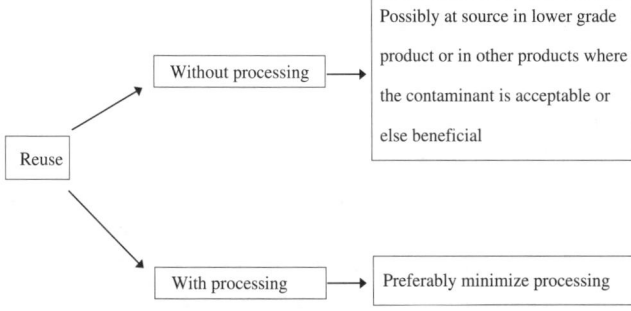

Figure 11.2   Recycling options.

The solvent recovery processor is usually concerned only with the reuse category, which can be further subdivided, as in Figure 11.2. Normally the solvent recycler is involved in reuse with processing, but occasionally receives waste which requires little or no processing and can be sold direct to a suitable end user. The reuse with processing category can be expanded to explore the option open for commercial recycling. A waste solvent can be reclaimed either for reuse as a solvent, or for its fuel value. Reuse as a solvent is usually preferable, but it is possible that other considerations based on a life cycle analysis would make recycling for fuel the preferred option.

Table 11.1 shows some of the options for recycling listed broadly in order of preference.

Solvent recovery is now inextricably linked with the alternative fuels market and developments such as Cemfuel (see Chapter 12), used as cement

TABLE 11.1
Solvent recycling with processing

- Minimal cold processing — for example, filtration, decanting, settling, blending
- Simple 'wet' distillation by direct steam injection
- Simple 'dry' distillation by indirect heating
- Simple batch fractional distillation
- Complex fractional distillation — high vacuum, continuous
- Simple fuel blending
- Complex fuel blending, dissolving, melting, macerating, etc

kiln fuel, have significantly changed the economics of recycling. The availability of a cost-effective way of treating distillation residues to convert them to a fuel means that wastes previously considered unsuitable for distillation because of the low yield and high waste disposal cost are now acceptable.

Critics of the Cemfuel programme have suggested that recyclable wastes might be included in the fuel — thus lowering their treatment down the hierarchy — whereas the reverse is true: more efficient use of the residues encourages recycling. It is clearly not in the interests of the recycler to dispose of a recycled solvent — which can be sold — into a fuel which at worst requires a nominal payment to the end user.

FINANCIAL CONSIDERATIONS

The possible options for solvent recycling have been examined, but to complete the picture the costs involved must be investigated.

It is often the case that after all the costs have been accounted for, the required selling price of the reclaimed solvent exceeds the virgin price and obviously recovery is not viable. In these cases the only way to 'balance the books' is to make a waste disposal charge to the producer.

In the case of toll recovery, or 'laundry', where the waste producer is also the customer for the recovered product, it is not unusual to expect to pay more for the laundered solvent than for virgin product. In this case the waste producer is essentially receiving free waste disposal but paying a premium for the recovered solvent.

Table 11.2 shows some of the principal costs. Some of the costs can be well controlled by using energy- and labour-efficient technology; others are relatively uncontrollable. Often the transport costs can be so high as to make recycling impossible for one operator, but very profitable for a processing facility located close to the producer of the waste.

The alternative fuels market and its effect on costs have already been mentioned. Further consideration shows that waste disposal costs — that is, the costs experienced by the recycling facility — have a dramatic effect on the viability of recovery. Toll processing of waste solvents for recycling has a yield factor; product recovered from a waste has a resale value. What remains is waste, and it incurs a disposal charge. There is a point at which the yield becomes too low and the waste disposal costs too high, and the waste becomes unrecoverable. If the cost of disposal of distillation residues is reduced, then lower yield waste can be accepted for recycling.

TABLE 11.2
Principal costs of recycling

1	Collection and transport to site
2	Energy for processing
3	Labour costs
4	Compliance costs
5	Maintenance
6	Waste disposal
7	Depreciation and capital investment
8	Product delivery

Solvent recovery is a Part A process authorized by the Environment Agency. This — coupled with health and safety requirements, fire risks and the strict controls relating to the collection, transport and treatment of wastes — means that compliance costs are high. In return, the 'duty of care' element of the Environmental Protection Act 1990 should mean that more producers will choose solvent recyclers with high standards, and be prepared to pay for those standards.

ASSESSMENT OF WASTE FOR PROCESSING AND RECOVERY STREAMS
The acceptance procedure for waste involves numerous considerations to decide if it can be processed and what is the best processing route. Often the waste producer specifies what must be done with the waste. Health, safety and environmental considerations may dictate how the waste is treated and, of course, the economics of any proposed treatment need to be acceptable.

All waste streams accepted by the waste processing and recycling company Solrec undergo a strenuous acceptance procedure to ensure the material can be viably processed. The procedures are quality assured to ISO9002 standards and non-conformance is rigorously investigated.

All material is accepted for processing on the basis that it conforms to the sample and information supplied by the waste producer.

The starting point is to obtain a sample with full health and safety data and to complete a questionnaire with the waste producer. The questionnaire prompts the waste producer to declare any significant information about the waste which could cause health and safety or environmental problems. Solrec

129

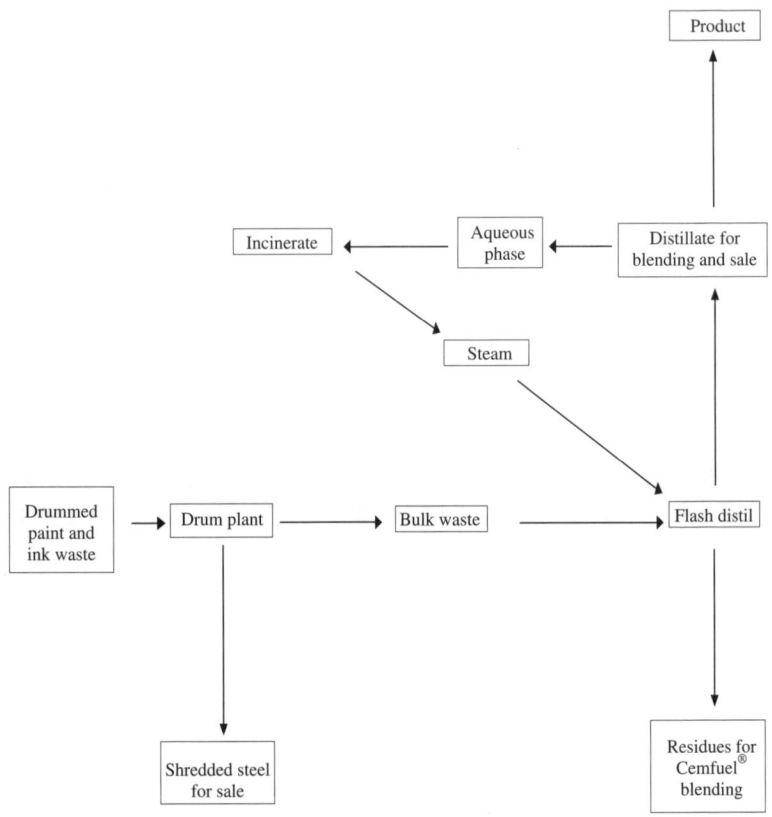

Figure 11.3 Flash distillation of high solid paint and ink wastes.

specifically excludes certain compounds such as polychlorinated biphenyls (PCBs), radioactive substances, biologically active materials, cyanides and so on, and requires written confirmation from the producer that the waste contains no excluded species.

The next stage is a laboratory assessment of the sample, giving sufficient information to determine the best recycling route.

If the proposed recycling route and costs are acceptable to the customer, then a plant trial takes place and, if this is satisfactory, the waste stream is accepted for regular processing. Solrec has four principal recycling routes; Figures 11.3, 11.4, 11.5 and 11.6 show the main stages in the processing. The processes are interdependent and combine to give an efficient processing route.

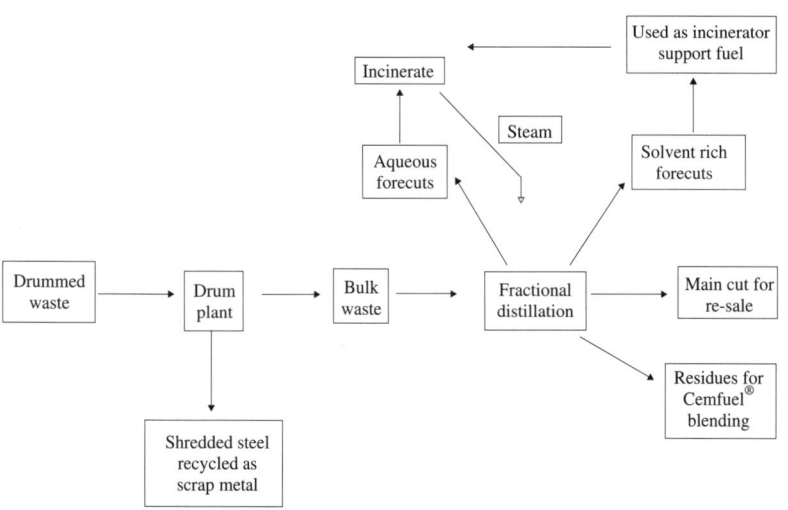

Figure 11.4 Fractional distillation.

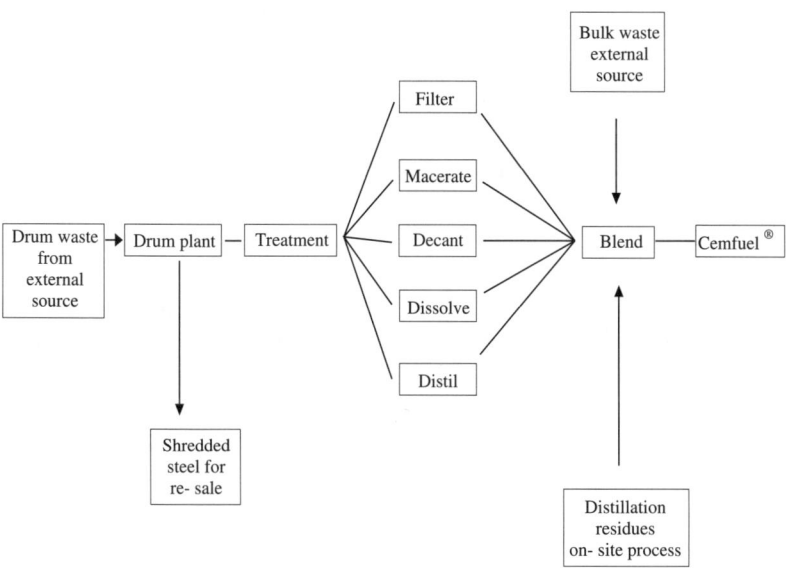

Figure 11.5   Cemfuel.

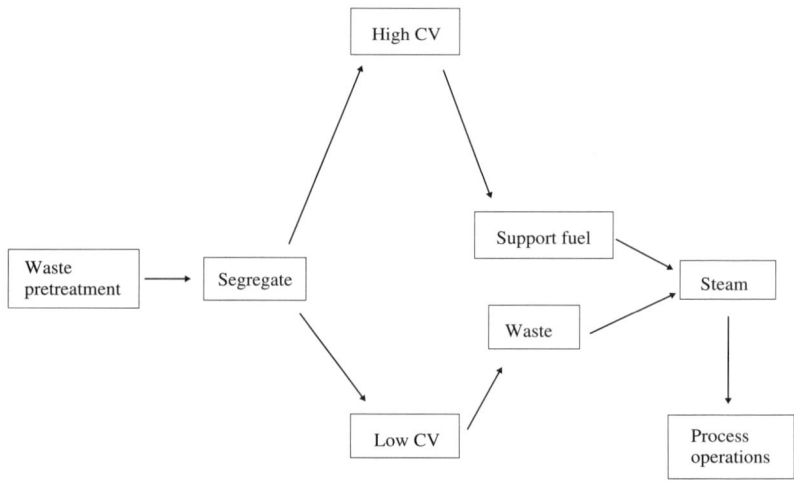

Figure 11.6   On-site combustion and steam generation.

For example, a waste stream containing 70% water and 30% organics and solvents has a high water content which precludes it from direct blending for Cemfuel; some of the organics and solvents contain chlorinated compounds and exclude the waste stream from on-site combustion and steam generation. The organic solvents, even if distilled, have no market.

Solrec would distil the waste stream to extract the water and some soluble solvents as a distillate. This water would be disposed of via the site combustion unit, which also raises steam to heat the distillation unit. The residues from the distillation, which are primarily organics and have a significant calorific value, are then blended to make Cemfuel.

Aqueous distillate burned on the combustion unit may contain 25% solvents which will not burn unsupported, but give a net positive heat release when burned with a support fuel.

CASE STUDY 1

PROBLEM
A car manufacturing company wanted to improve its environmental and waste disposal performance with regard to solvent usage in the spray lines.

At the start of the project, no recycling was carried out and both solvent waste and aqueous waste were directed to a common storage tank for disposal by incineration. Every car that travelled down the line was a different colour and between each car the paint spray line would be flushed clear with a virgin solvent blend.

This example contains some of the basic elements common to most recycling operations; the solution required investment and process change.

SOLUTIONS

Segregation

In this case the aqueous waste was mixed with the solvent waste. The mixture, if offered for recovery, would have incurred high processing costs and resulted in low yield and poor quality recovered solvent. If the waste were incinerated, then once again high charges could result because of the low calorific value.

By segregating the wastes, the solvent waste was much more suitable for recovery and the aqueous wastes could be treated by conventional effluent treatment plants.

To achieve segregation required additional plumbing and tanks, and modifications to the spray nozzle flushing systems.

Recycling solvent waste distillate as the flush solvent

This proposal seems an obvious improvement, but is practically much more difficult to achieve. Solvents recovered from waste streams are variable in their specification, and the nature of quality-controlled production in the motor industry demands a consistent raw material.

In this case, it was necessary to determine the factors which were important in the specification of the recovered solvent, as it proved technically impossible to recycle the waste to the same specification as the virgin flush solvent.

Two factors were significant. First, the solvency of the recovered solvent had to be adequate to dissolve the paint and resins in the spray lines. Second, the conductivity of the recovered solvent had to be correct to be compatible with the spraying process equipment.

Solrec was able to take the recovered solvent from the spray process and, by the addition of small quantities of virgin solvents, meet both the solvency and conductivity requirements of the car manufacturer.

133

CONCLUSION

Using this solution, the manufacturer was able to recycle 1000 tonnes per annum of waste solvent and dramatically reduce waste disposal and virgin solvent costs.

CASE STUDY 2

PROBLEM

Solrec wanted to improve its waste disposal performance and attract additional recycling business.

The Environmental Protection Act 1990 introduced the concept of 'duty of care', whereby a waste producer must ensure that the chosen contractor for the disposal of the waste acts responsibly and in accordance with all regulations, licence authorizations, consents and so on.

Solrec's solvent recovery process produced distillation residues which were solidified and disposed of to a suitable licensed landfill site. The disposal route was entirely legal and met all environmental requirements, but many waste producers — particularly American-owned subsidiaries — would allow their wastes to be recycled only if the residues were destroyed by incineration. The cost of commercial incineration was too high to make recovery viable and often the entire stream would be incinerated.

SOLUTION

Solrec had long been aware that distillation residues had an excellent energy value as a fuel, but the combustion of the fuel would require a very special end user. Castle Cement knew that its cement kilns operated at extremely high temperatures, long residence times and high turbulence, with excellent flue gas back end equipment — all factors which would be required to use distillation residues as a fuel.

Castle Cement and Solrec started discussions in 1989. Castle Cement was able to produce a required fuel specification based on three fundamental principles. The fuel must not:

• harm the environment;
• endanger the health and safety of employees;
• affect the quality of the cement adversely.

These fundamental parameters resulted in a very detailed fuel specification and Solrec took up the challenge to alter the recovery process to allow residues to be used as fuel.

Distillation residues from the solvent recovery process were too viscous to be handled by the cement works and contained solids such as rags, wood, plastic and polymerized resin. They also contained some chemical components which were not compatible with the three fundamental parameters. Additionally, chemical analysis of the residues required very sophisticated analytical equipment and considerable method development.

To solve the problems, Solrec first acquired additional thin waste solvent streams which were of a type not normally recyclable. These were used to reduce the viscosity of the residues. Secondly, investment was made in macerating and solids reduction equipment to process the solids in the waste to a size suitable for pumping. Thirdly, investment was made in an integrally coupled plasma spectrometer (ICP), calorimetric equipment and ion chromatography equipment to ensure that non-compatible residues were identified and removed from the Cemfuel manufacturing process.

Considerable investment was made by Castle Cement to enable the fuel to be used and in 1992 Solrec made the first delivery of Cemfuel, a highly specified pumpable liquid fuel tailored to be compatible with cement manufacture and able to replace up to 50% of the energy requirements of the kiln.

CONCLUSION

Solrec found an environmentally more favourable solution for residue disposal, which met with the approval of numerous major waste producers and allowed more solvent recovery. Castle Cement found a more cost-effective fuel which reduced emissions and produced a better quality cement. Both parties had to undertake considerable development and capital costs.

Solrec now processes around 5000 tonnes per year of distillation residues into Cemfuel, which replaces the previous practice of landfill and elevates this volume of waste up the disposal hierarchy to that of beneficial reuse.

12. CEMFUEL — A FUEL FOR CEMENT KILNS

Iain Walpole

INTRODUCTION

Portland Cement is a key constituent of concrete, representing 10–15% by mass of this ubiquitous and environmentally-friendly construction material. The manufacture of cement is essentially a three stage process — preparation of raw materials, clinkering and finish grinding. The production of cement clinker is the heart of the process and is carried out in rotary kilns. The chemical reactions required to produce clinker can only occur at high temperatures and require long residence times.

Cement kilns are very large machines which use enormous quantities of energy in the production of a quality product. The choice of fuel used by the manufacturer represents one of the most significant costs. Cemfuel is a means of dramatically reducing these energy costs by utilizing the high temperatures, long residence times and high thermal inertia within the cement kiln — attributes required for cement production.

Cemfuel is a highly specified liquid fuel, manufactured from certain selected waste streams. Its real claim to fame, though, is that it reduces the environmental impact of the cement manufacturing process when compared with normal coal firing. The use of Cemfuel has been shown to reduce global emissions of 'greenhouse' and 'acid rain' gases.

The recent White Paper *Making Waste Work*[1] sets out the UK Government's strategy for sustainable waste management. The Paper recognizes the role cement kilns can play in moving waste up the hierarchy from disposal to energy recovery. Paragraph 2.165 states:

'The government takes the view that:
- the use of cement kilns ... to destroy wastes is a valuable way of recovering energy from them provided that emissions are tightly monitored and controlled, so as to protect human health and the environment; and
- for some waste streams energy recovery from cement kilns may be the BPEO and could play an important role in meeting recovery targets.'

This chapter sets out to demonstrate that the use of Cemfuel satisfies these points and to show that the use of Cemfuel is part of the best practicable environmental option (BPEO) for cement manufacture.

BACKGROUND

Over the past 30 years the fuel requirement per tonne of cement has been reduced by 40% through investment in new dry process kiln systems and using state-of-the-art technology[2]. Further technological improvements will yield only marginal savings on unit costs. Fixed costs have been aggressively reduced by plant closures and rationalization.

In order to remain competitive in an international situation, where imports sold at the margin control the selling price of cement, Castle Cement (the second largest cement company in the UK) needed to examine closely its energy costs. Electricity costs in the UK are considerably higher than those of continental competitors and cannot be controlled, so kiln fuel costs — which still comprise 40% of the variable cost of production — needed attention.

Castle Cement's annual kiln fuel consumption includes some 400,000 tonnes of coal and a small quantity of gas oil. Castle Cement researched all the options and concluded that the recovery of energy from waste was the best practicable way forward.

MANUFACTURING CEMENT

Cement is manufactured from a group of minerals taken from local quarries which provide the necessary silica, calcium, iron and alumina. The object of the process is to combine the calcium and silica to form calcium silicates. This reaction can only take place at a temperature of over 1450°C and in a flux formed by the iron and alumina components of the kiln feed. In a cement kiln, the temperature of the gas stream reaches over 2000°C and is maintained at this temperature for a residence time of several seconds. These process conditions are considerably hotter and more prolonged than those found in the best incinerators.

The atmosphere in the kiln is very turbulent and highly alkaline. It should be noted that the ash from any fuel used is absorbed into the raw material stream and becomes part of the product which, in its final concrete form, is an extremely stable inert material.

THE CEMFUEL PROGRAMME

Cemfuel was designed to take advantage of the high temperatures, long residence times and highly turbulent alkaline atmosphere of cement kilns. In order to minimize the project development and lead-in time, specialist knowledge of solvent recovery and organic chemistry was required. A co-operative agreement was signed with Solrec Limited, of Heysham, Lancashire, UK, who already had the complementary expertise and knowledge.

The specification — the most fundamental part of the Cemfuel programme — was developed to allow a range of organic materials from a wide variety of industries to be processed into a high specification cement kiln fuel. Chemicals such as alcohols, paraffins, ketones, olefins and organic acids provide the base, while exclusions include biologically active materials, radioactive materials and clinical wastes. Materials that could have health and safety implications for Castle Cement's employees or neighbours, as well as all those

A selection of products from which Cemfuel may be derived.

chemicals which would have an impact on emissions or cement quality, are restricted or excluded. It is because of this tight specification, which also covers the manufacturing processes and quality control regimes, that Castle Cement is satisfied that Cemfuel is designated a fuel in accordance with the EU framework directive on waste[3].

EMISSIONS MONITORING

Castle Cement embarked on an extensive environmental monitoring programme for its Cemfuel-fuelled kilns. A protocol was developed, requiring baseline checks using conventional fuels before each analytical run, to take into account some of the natural variations in raw materials and coals. In selecting the various analytical and sampling methods, Castle Cement and its consultants reviewed many different techniques from around the world.

At the start of the Cemfuel programme in 1991 the UK and Europe had not developed satisfactory standard methods for monitoring a number of stack emissions; for many emissions this position remains unchanged. An independent test team was chosen to conduct the monitoring work. As the studies progressed the methods were changed and updated. For example, spot sample results for sulphur dioxide (SO_2) and oxides of nitrogen (NO_x) were found to be unrepresentative, so continuous monitoring was adopted for these gases.

The comprehensive emissions study on Castle Cement's kilns demonstrated that using Cemfuel at replacement levels up to 50% reduces SO_2, NO_x and halide emissions, and that there are no changes in emissions of trace metals, dioxins, furans and total organic carbon (TOC) when compared with normal coal burning.

Castle Cement has shown that emissions of organic compounds, such as dioxins, are totally independent of the fuel being used, as a result of the high combustion temperatures. The very small quantities found in cement kiln exhausts are formed from precursors in the raw feed. This is demonstrated by examination of the congener profiles[4]. The same is true of other hydrocarbons, expressed as TOC emission, which originate from organic components from within the raw materials for cement-making rather than from the fuels.

The reduction of NO_x is attributable to the change in flame characteristics as the volatile solvent streams are added to centre of the coal flame. The flame front temperature is reduced and less atmospheric nitrogen is fixed.

The coal or pet-coke normally used as fuel are comparatively high in sulphur. Sulphur inputs to the kiln are reduced by substituting Cemfuel for these

The Cemfuel reception tanks at the Ribblesdale works.

higher-sulphur fuels. However, it is important to bear in mind the strong scrubbing action of the alkaline conditions within the cement kiln and the preheater tower which removes most of the fuel sulphur. Indeed, the preheater tower on dry process kilns operates as an extremely effective dry scrubber. Test work has demonstrated that all the fuel sulphur entering the preheater from the kiln and calciner is absorbed by the free lime from the raw materials in the lower stage cyclones.

SO_2 emissions from a cement kiln are largely a result of pyrites in the raw feed. These compounds are burnt off in the upper stages of the preheater so there is limited opportunity for the SO_2 to combine with the lime from the raw materials. This explains why there was no reduction in SO_2 emissions at Castle Cement's Ketton works, despite the reduced sulphur input to the kiln. Castle Cement found a larger reduction in SO_2 emissions than expected on wet process kilns at its Ribblesdale site, probably as a result of changes in the sulphur/chlorine balance.

The metals in a cement kiln react in either a volatile or a refractory manner. The volatile metals leave the kiln burning zone as vapour and condense in the cooler regions of the plant. The refractory metals react in the same manner as the cement-making metals and become part of the crystalline structure of the cement minerals. The volatile metals are tightly restricted in the Cemfuel speci-fication (for example, mercury is restricted to less than 20 mg kg^{-1}) so it is not surprising that they have no effect. Recent studies carried out in Belgium indi-cate that the capture of volatile metals such mercury can be as high as 90% (that is, most of the mercury entering the kiln is trapped in the clinker or in the kiln dust).

Overall, the input of environmentally sensitive metals is overwhelm-ingly from the raw materials and so no effect due to fuels is measurable. This has been demonstrated by the numerous stack emissions tests completed by Castle Cement[4].

Chloride emissions from cement kilns are normally in the form of potassium chloride although it is the chemist's convention to express these emissions as HCl. Any chlorine in the fuel to the cement kiln is removed from the fuel molecules in the high temperature zone and reacts to form a common salt in the cooler parts of the kiln system. The normal salt formed is potassium chloride with any residual chlorine mopped up by sodium or calcium in accord-ance with classical chemistry.

Measurements of chloride emission, expressed as HCl, show a de-crease when burning Cemfuel, probably as a result of the change in the ratio of chloride to sulphate in the kiln charge.

Finally, Cemfuel reduces carbon dioxide (CO_2) emissions at the ce-ment plant when compared to the use of coal because of its higher hydrogen to carbon ratio.

ENVIRONMENTAL MONITORING

Before Cemfuel was introduced at Ribblesdale and Ketton, a baseline study of the local environment was carried out. This included soil sampling from a num-ber of locations selected from plume dispersion modelling. These soil samples were analysed for dioxins, furans and trace elements. The baseline surveys included ecological surveys of the nearby rivers, including a phase 1 habitat sur-vey[5] and inventory of flora in the river corridors. Ambient air quality was checked using the moss bag technique.

These baseline studies grew into a comprehensive environmental monitoring programme that Castle Cement's consultants carry out twice yearly at all the company's clinker-producing factories. More recently ambient SO_2 and NO_2 have been measured using diffusion tubes; this data complements that collected for particulates over many years using dust deposition gauges. The latest addition to Castle Cement's environmental monitoring programme is the study of air quality effects on flora. Whilst this work is in its infancy, early results suggest that the impact of Castle Cement's work on the local environment is very small indeed.

THE BEST PRACTICABLE ENVIRONMENTAL OPTION

In November 1993 Her Majesty's Inspectorate of Pollution (HMIP) issued the Integrated Pollution Control (IPC) authorization for Castle Cement's Ribblesdale works. Included in the authorization was the use of Cemfuel. Following a sustained period of public and political pressure, in September 1994 the inspectorate issued a variation notice that required Castle Cement to carry out further emissions monitoring work and a full best available techniques not entailing excessive cost (BATNEEC) study to justify the continued use of Cemfuel[6].

In consultation with HMIP it was agreed that the most appropriate way forward in the absence of any other methodologies was to use HMIP's own draft BPEO assessment procedure[7]. The BPEO assessment of Cemfuel was carried out by AEA Technology[8]. This study was probably the first to use the guidance outside the pilot studies. A similar study was carried out as part of the application for permanent use of Cemfuel at the Ketton works[9].

The central control room at the Ribblesdale works.

The Cemfuel programme has many environmental benefits, both at the cement works and globally. The manufacture of Cemfuel and its use as a cement kiln fuel takes waste out of landfill and incineration. By burning waste-derived fuels in cement kilns the energy content of organic wastes is recovered and used in the manufacture of Portland Cement. This recovery of energy also has the effect of a reduction in fossil fuel consumption and a reduction in CO_2 emissions.

The HMIP BPEO assessment methodology has passed through a number of drafts in the last two years, but common to each stage of the development is the restriction of BPEO to site-specific issues — so many of the global benefits of the Cemfuel programme actually fall outside the scope of the study. This restriction to the scope of the technique was due to the constraints placed on HMIP under the Environmental Protection Act 1990. The formation of the Environment Agency in April 1996 and its responsibility for waste strategy may change this narrow definition of BPEO.

THE ENVIRONMENTAL BALANCE SHEET

It is Castle Cement's view that the global environmental benefits of the Cemfuel programme should be taken into consideration in any BPEO assessment. In truth, some of these environmental benefits are difficult to quantify and perhaps the most appropriate way to fully understand the effects would be by undertaking a life cycle analysis for each waste stream entering the Cemfuel programme. This would give the BPEO for the disposal of waste, but not necessarily the manufacture of cement.

As well as the environmental benefits of the use of Cemfuel, there is a reduction in the power requirements for the preparation of pulverized fuel. The power consumption per tonne of fuel burnt is reduced by replacing solid fuel handling systems with small pumps. This reduction in power consumption leads ultimately to a saving in emissions at the power station.

By replacing coal with Cemfuel there is a global reduction in CO_2 emissions (assuming the components of Cemfuel would normally have been incinerated without energy recovery). Table 12.1 shows an 'environmental balance sheet' over the period of one year for a single cement kiln replacing 40% of kiln fuel with Cemfuel. Inspection of this table shows the strength of the case that Cemfuel reduces the environmental impact of cement manufacture.

THE CEMFUEL BPEO ASSESSMENT

The BPEO assessment carried out for Castle Cement concentrated on the impact

TABLE 12.1

The environmental balance sheet of the Cemfuel programme

	Units	Base case	Cemfuel case
Waste to landfill or incineration without energy recovery	Tonnes	28,000	0
Fossil fuel used in kiln	Tonnes	60,000	36,000
Clinker produced	Tonnes	300,000	300,000
Global CO_2 released from fuel only	Tonnes	253,000	182,600
Power consumption, fuel preparation and handling	MWh	3000	2000
Fuel consumption for coal drying	Tonnes	36	21
CO_2 from power generation	Tonnes	3045	2030
Global warming potential	Tonnes CO_2	256,177	184,709

of the clinker manufacturing process. Prior to the study it was Castle Cement's view that the use of Cemfuel left most emissions from the cement kiln unchanged except for a small reduction in SO_2 and an appreciable reduction in NO_x. Therefore Castle Cement expected that the BPEO methodology would result in two very similar integrated environmental indices.

The study itself considered releases to air and waste arisings. Releases to water from the cement industry are restricted to small amounts of cooling water, so the change of fuel would not change the already negligible impact of the releases to surface water.

The BPEO assessment demonstrated that the most significant emissions from cement kilns are SO_2 and NO_x, both of which have been reduced following the introduction of Cemfuel.

The overall value of these studies must be put into context with errors arising at the various stages of the assessment procedure — for example, stack monitoring data are typically 50% or more and the errors inherent in plume dispersion modelling are around 20%. Several rather imprecise numbers have thus been combined to give a single figure, the integrated environmental index (IEI), to characterize the environmental impact of the process.

A further area of difficulty arises in the interpretation of emissions monitoring data. Many of the emissions measured at both Ribblesdale and Ketton have been below the limit of detection. Considering trace metals, the limits of detection of analytical equipment vary depending on the size of the sample and how the instrument is set up. For example, a laboratory may analyse one metal regularly, so the instrument will have a very high resolution for this element, whereas another metal might only be tested occasionally and the resolution will be coarser. The net effect of a high limit of detection on the environmental index is most apparent for elements with very low environmental assessment levels (EAL). The element with the lowest EAL is cadmium. Even when possible errors in analysis are considered, the overall effect of the resolution of analysis techniques has an insignificant impact on the IEI.

There is a secondary effect to be considered. Throughout the development of the BPEO methodology HMIP recognized that EALs may change from time to time as a result of new knowledge. The number of compounds with EALs will increase, so that rather than a single figure covering all nickel compounds there will be EALs for, say, the oxide, the chloride, the sulphate and a range of organo-nickel compounds. Fortunately for the cement industry, the major pollutants — particulates, NO_x and SO_2 (almost 95% of the IEI) — are covered by statutory environmental quality standards, so there are unlikely to be any radical changes to the IEI for a cement kiln as a result of changing EALs.

Considering trace metals, such a change has already occurred. Between the Ribblesdale study and the Ketton study the EALs for a number of metals were revised. For example, the 1994 EAL for nickel was 0.0025 $\mu g \ m^{-3}$ which was then revised to 0.2 $\mu g \ m^{-3}$ in 1995. Looking at the 5% of the IEI made up of the effect of metals, at Ribblesdale nickel represents 74% of the IEI for metals using the 1994 figure and 0.4% using the 1995 EAL.

The results of the Castle Cement BPEO assessments are summarized in Table 12.2. The IEI is not intended for comparison between different processes. However, recently published worked examples[10] give IEI values that are two orders of magnitude larger than those given in Table 12.2. Clearly, compared with those processes, cement manufacture at the Castle sites makes a small contribution to pollutant concentrations in the environment.

From Table 12.2 it can be concluded that the environmental impact of the cement manufacturing process is insignificant and that the change of kiln fuels has an insignificant effect on that impact. Interestingly, the mass emissions of most pollutants are lower at Ketton than Ribblesdale, but the BPEO methodology suggests Ketton has the higher IEI. This difference is the result of better dispersion at Ribblesdale due to higher stack emission temperatures.

THE FUTURE

The UK cement industry burns approximately 2 million tonnes of fuel per year. Castle Cement believes that 50% replacement of conventional fuel is an achievable target. Across the UK cement industry this represents a saving of 1 million tonnes and a substantial contribution to Government waste recovery targets. The potential for fuel replacement does not stop at Cemfuel; Castle Cement is already working on the use of tyres and tyre chips at Ketton.

The EU Packaging Waste Directive places recovery targets on the recycling and reuse of packaging throughout the European Union. Paper and plastic packaging that cannot be reused could also be used as a cement kiln fuel. This would help industry to achieve the mandatory packaging recovery targets of 50% to 65% of all packaging.

Other materials which have been used by the European cement industry as alternative fuels include waste oils, paper, board, wood, plastics and sewage sludge. During 1995 the Swiss cement industry formally agreed with the federal government to replace 75% of fossil fuels with waste-derived fuels.

Finally, the manufacture of cement requires raw materials. Certain selected waste streams can be used as a source of calcium, silica, alumina and iron oxide. The industry has been utilizing materials such as pulverized fuel ash,

TABLE 12.2
Summary of BPEO assessment

	Ribblesdale		Ketton	
Fuel mix	Coal only	Coal + Cemfuel	Coal only	Coal + Cemfuel
IEI	0.019	0.014	0.027	0.025
Global warming potential	8.6	7.5	8.7	8.4
Ozone generation potential	Equal	Equal	Equal	Equal
Waste hazard index	15,400	15,700	9990	9700

Kiln 7 at the Ribblesdale works.

mill scale and slags for many years. There is the potential to increase the utilization of other wastes such as spent catalysts, providing there are no detrimental effects on product quality or the environment.

CONCLUSIONS

The use of the HMIP BPEO assessment methodology demonstrates that the use of Cemfuel represents the site-specific BPEO for the manufacture of cement at both Ribblesdale and Ketton. In all cases, irrespective of kiln fuel mix or raw material mineralogy, the IEIs are extremely low. The emissions of SO_2 and NO_x are the dominant contributors to the environmental index and both these emissions are reduced by the use of Cemfuel.

The environmental balance demonstrates that the use of Cemfuel has global environmental benefits, including reducing CO_2 emissions, reducing the consumption of non-renewable fossil fuels and reducing the use of landfill and incineration for waste disposal.

The use of Cemfuel moves materials up the waste hierarchy from disposal to recycling by energy recovery in accordance with the Government waste strategy. This demonstrates the valuable role the cement industry can play in sustainable waste management.

Castle Cement is satisfied that a waste generator's 'duty of care' responsibilities are fulfilled when the waste enters the Cemfuel process. Even so, the company welcomes duty of care visits and provides waste generators an open book on the environmental impact so that they can satisfy themselves that Cemfuel is the BPEO.

REFERENCES IN CHAPTER 12

1. Department of the Environment and the Welsh Office, 1995, *Making Waste Work, A Strategy for Sustainable Waste Management in England and Wales* (HMSO).
2. British Cement Association, 1994, *UK Cement Manufacture and the Environment* (British Cement Association).
3. Framework Directive on Waste 75/442/EEC.
4. *Commentary on BPEO Report and Associated Tests on Kiln 5 at Ribblesdale Cement Works* (Castle Cement).
5. NCC (Nature Conservancy Council), 1989, *Phase 1 Habitat Assessment Handbook.*
6. Ribblesdale Works IPC Authorization Variation AN7431, September 1994.
7. *Draft Technical Guidance Note E1, Environmental, Economic and BPEO Assessment Principles for Integrated Pollution Control*, Volume 1 (HMIP).
8. BPEO Assessment of the Burning of Cemfuel at the Ribblesdale Works of Castle Cement Limited, AEA/CS/RPRO/16411099.
9. BPEO Assessment of the Burning of Cemfuel at the Ketton Works of Castle Cement Limited, AEA/RPRO/16411128.
10. *Draft Technical Guidance Note E1, Environmental, Economic and BPEO Assessment Principles for Integrated Pollution Control*, Volume 2 (HMIP).

13. THE INCINERATION OF SOLVENTS AND OTHER HAZARDOUS WASTES

Tony Dean

HISTORICAL PERSPECTIVE

The hazardous waste incineration industry in the UK was created and nurtured by consecutive tranches of environmental legislation in the last 25 years, starting with the Deposit of Poisonous Substances Act 1972. The Act was the first to insist that a defined list of poisonous substances or agents must be treated to remove their human or ecotoxicity prior to disposal via landfill or watercourse.

This was followed, on a wave of public interest in environmental matters, by the Control of Pollution Act 1974 and its consequent regulations. These introduced the concept of 'Special Waste', and the '20 kg child' test. All disposals of Special Waste from industrial or commercial premises had to be documented 'from cradle to grave' and be disposed of at specially licensed premises. In general the disposal site licences defined the nature of the site and the criteria by which individual waste streams could be accepted for disposal or treatment at those sites.

This system worked well and purpose-built hazardous waste incinerators were the best practicable environmental option for the treatment of many of the more complex chemical and solvent wastes. In particular, these included chlorinated solvents, pharmaceutical process wastes and by-products, agrochemical process wastes and other complex wastes resulting from general chemical and polymer manufacture. Her Majesty's Inspectorate of Pollution (HMIP) controlled the emissions to air from incinerators using 'best practicable means' as a philosophy.

The Environmental Protection Act (EPA) 1990 followed some years later, and included waste incineration and waste solvent recovery by distillation in its Part I processes, resulting in further tightening of the regulations governing the operation of purpose-built incinerators.

Public fear of the potential for pollution incidents in the handling of the most hazardous chemical wastes has meant that the regulations applied to hazardous waste incinerators have been the tightest of all the waste disposal processes, and latterly of all the industrial processes controlled by Integrated Pollution Control (IPC) under Section 1 of EPA 1990. This Act took control of the operations of a waste incinerator from the local authorities' waste regulators for waste movements, acceptance, storage and emissions to land, HMIP for air emissions and the National Rivers Authority (NRA) for emissions to water, and gave HMIP total control of all of these issues. On 1 April 1996 this control passed, with HMIP, to the Environment Agency. The change to IPC (in 1993 for incineration processes) was the vehicle by which further tighter emission limits were imposed on incinerators due to EPA 1990.

The final piece of relevant legislation is the EC Hazardous Waste Incineration Directive (HWID) which came into force in December 1994, once again increasing the pressure on incinerators to reduce emissions to those practically achievable.

These ever-tightening sets of legislation were driven by public concerns over waste incineration. In Europe the regulations were based on 'best available techniques' — the 'not entailing excessive cost' part being dropped in the case of hazardous waste incineration. This has created a high-tech industry whose environmental impact is as minimal as technology allows. These plants are costly to build (approximately £50M for a 60,000 tonne/annum plant), carry very high fixed costs, and make hazardous waste incineration expensive to the waste producers. This latter fact has had a beneficial effect in encouraging waste minimization within the pharmaceutical and agrochemical industries.

HAZARDOUS WASTE INCINERATOR DESIGN IN THE 1990s (EXAMPLE — CLEANAWAY'S HWI AT ELLESMERE PORT)

The first step in any incinerator design is to ensure safe waste storage and feeding. The second step is to ensure careful and thorough design of the combustion chambers to ensure maximum destruction of the hazardous molecules and safe entrapment of the resulting non-volatile pollutants. The third step is to design a flue gas cleaning plant capable of meeting the tightest pollution emission limits to water, land and air of any industrial process.

At Cleanaway's Ellesmere Port hazardous waste incinerator (HWI), drummed solid materials are kept in a purpose-built drum store fitted with air extraction and a foam sprinkler system (see Figure 13.1). Liquids are stored in mild and stainless steel storage tanks with full nitrogen blankets. Exhaust gases from the tanks, made up of nitrogen and various volatile impurities, are fed into the incinerator. When the incinerator is shut down for maintenance, the vapours pass through a carbon filter.

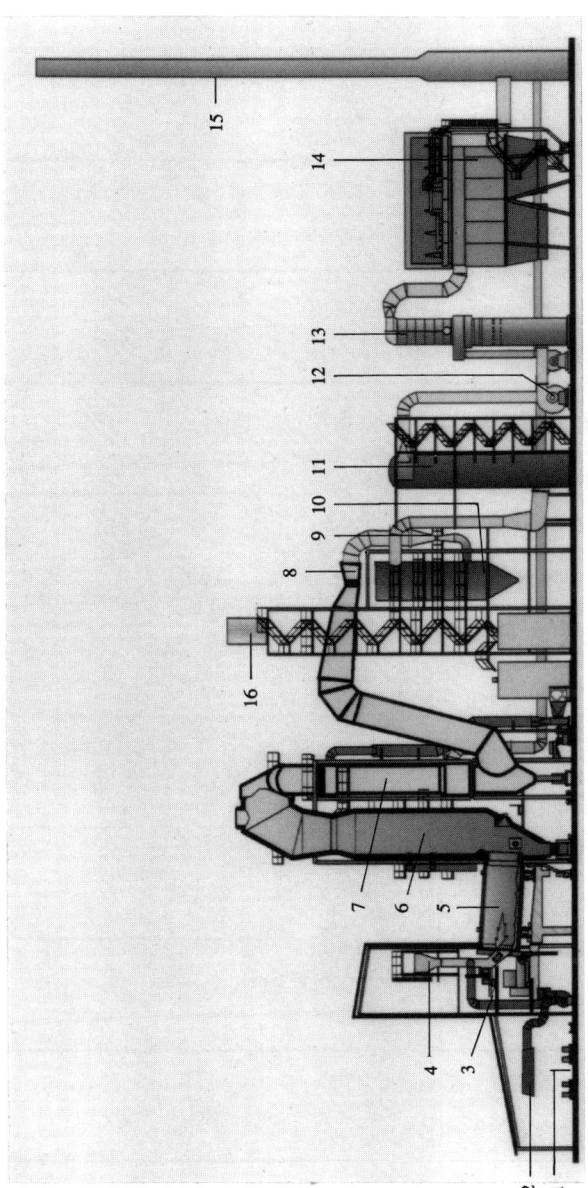

Figure 13.1   Cleanaway's high temperature incinerator, Ellesmere Port, Cheshire, UK.

1 — packaged waste conveyors; 2 — fume extraction; 3 — packaged waste hoist; 4 — bulk solids hopper; 5 — rotary kiln; 6 — secondary combustion chamber; 7 — gas-gas heat exchangers; 8 — saturate venturi; 9 — variable throat venturi scrubber; 10 — first scrubbing tower; 11 — second scrubbing tower; 12 — induced draught fan; 13 — hot air mixer mist eliminator; 14 — fabric filter; 15 — stack; 16 — water tank.

The combustion equipment is in two sections. The first unit is the rotary kiln, kept at a constant temperature by burning high calorific value liquid wastes. Drummed solid and liquid materials are also burned in the kiln, with the drum also being consumed. The kiln is refractory lined and is fed with sand and glass to create a protective slag layer. The slag encapsulates metal-containing ash in a vitrified bead which is non-leachable, therefore minimizing emissions to the environment when landfilled. Part-burned vapours from the kiln are passed into the secondary combustion chamber where further liquid wastes are

Cleanaway's high temperature incinerator, Ellesmere Port, Cheshire, UK

burned via main burners and lances. The lances and secondary air ports are arranged tangentially round the cylindrical structure to create a powerful vortex which ensures thorough mixing and complete burnout of vapours and solid particles. Residence times are over four seconds after the last waste injection point. This design is probably the best currently operating in the world and achieves a destruction efficiency for PCBs (one of the most refractory hazardous wastes) of 99.999996%.

The combustion gases are cooled to 750°C in two recuperators fabricated from Inconel alloy, producing hot air which is recycled to dry the flue gases later in the process. The gases are then instantaneously shock cooled to 80°C in a saturate venturi to prevent re-formation of dioxins. Two scrubbers are used to wash out acid gases with water and caustic soda and, when required, to wash out bromine with sodium sulphite solution. Having passed through the main induced draught fan, the wet flue gases are then re-heated by mixing with hot air from the recuperators to 95°C prior to final dust filtration in a fabric filter. The cleaned gases pass via a continuous emission monitoring station, where levels of hydrocarbons, CO, O_2, HCl, SO_2 and dust are measured, to the 80 m stack. The gas cleaning plant enabled the incinerator to meet the emission limits specified in the EC HWID, six years before the required date of 30 June 2000.

The acid solutions from the scrubber, entraining some particulate ashes, pass to the effluent treatment plant where lime addition, flocculation and settling tanks enable the water leaving the plant for the Mersey estuary to be generally cleaner than the river water taken into the plant. The facility to add special chelating compounds for precipitating difficult metals is available. In this way emissions to water are minimized. The centrifuged lime sludge is landfilled at a properly engineered and licensed site nearby.

THE RISE OF CO-INCINERATION

Co-incineration is the burning of hazardous wastes alongside other fuels in a process designed for purposes other than toxic waste destruction.

In the UK, co-incineration is confined to cement kilns, which have been burning increasing quantities of hazardous waste since 1992, using the combustion energy contained within the hazardous wastes to save money on conventional fuel.

The original waste disposal route via cement kilns was simple in construction. A waste management company, usually one of the smaller players, collected liquid and solid hazardous wastes from customers and mixed them in a stirred tank. As long as the mixture satisfied calorific value, ash, water, maximum PCB and heavy metal concentration specifications laid down by the cement companies, it could then leave the mixing station as a 'fuel' and be co-incinerated in the cement kilns. Recently, however, the Department of the Environment (DoE) and the Environment Agency have required that these materials remain a 'waste' all the way to the cement kiln. This definition as a waste is important in terms of the regulatory paperwork required to ensure traceable disposal routes which provide public information on waste producers' environmental performance, and ensures that the 'duty of care' of waste producers can be audited.

The cement-making process is designed to make clinker for cement manufacture. Cement kilns were not designed to produce high destruction rate efficiencies of hazardous wastes, and have relatively low flue gas cleaning efficiencies.

Under the IPC regulations produced under Part 1 of EPA 1990, cement kilns are generically governed by guidance note IPR 3/1, as opposed to IPR 5/1 which is the equivalent for high temperature incineration. Table 13.1 gives the comparative minimum requirements laid down by these two guidance notes, in terms of emission limits for known pollutants in the stack gases.

A comparison of the generic minimum requirements contained in IPR 5/1 (for high temperature incineration) with those in IPR 3/1 (for cement kilns) shows that IPR 5/1 applies emission limits to more pollutants (due to the variable composition of hazardous waste compared to fossil fuels), and that emission limits are generally lower. The compliance dates for existing plant are also earlier for incinerators.

It is from these regulatory starting points — produced for different processes operating in different markets with different materials — that the Environment Agency (formerly HMIP) has had to try and produce a set of rules for cement kilns burning hazardous waste that protects the environment as effectively as using purpose-built high temperature incinerators. The results, in terms of the individual authorized emission limits applied to cement kiln co-incinerators, are those given in Table 13.2 on pages 158–159.

The difference in emission limit controls can be traced to the nature of the cement clinker process. The raw material often contains significant sulphur levels and the high temperature coal flame produces more NO_x than incinerators. The calcination process by its nature is very dusty. In producing the individual authorization's emission limits, HMIP took into account what the existing cement kilns with simple electrostatic precipitator pollution abatement

TABLE 13.1
Guidance Note emission limits

Parameter units	Guidance for hazardous waste incinerators under EPA 1990 (IPR 5/1)	Guidance for cement kiln processes under EPA 1990 (IPR 3/1)
	New plant or by 1996 95% of hourly averages in 24 hours	New plant or by 1998 95% of hourly averages in 24 hours
Particulates, mg m^{-3}	20	50
Total organic carbon, mg m^{-3}	20	–
Hydrogen chloride, mg m^{-3}	10	–
Hydrogen fluoride, mg m^{-3}	2	–
Oxides of sulphur (as SO_2), mg m^{-3}	50	750
Oxides of nitrogen (as NO_2), mg m^{-3}	350	1500
Carbon monoxide, mg m^{-3}	50	–
Cadmium and thallium, mg m^{-3}	0.1	–
Mercury, mg m^{-3}	0.1	–
Sb, As, Pb, Cr, Co, Cu, Mn, Ni, V, Sn, mg m^{-3}	1.0	–
Dioxin TEQ (guide), ng m^{-3}	0.1	–
Dioxin TEQ (limit), ng m^{-3}	1.0	–

equipment could actually achieve (BATNEEC principle, compared with BAT for incinerators).

This approach allows hazardous waste materials to be burned in cement kilns, as long as the combustion of those waste materials does not produce more pollution than the coal and raw materials. However, the existing cement kiln processes cannot be as tightly controlled as high temperature incinerators in terms of emissions. HMIP therefore adopted control of pollutant levels in the waste input to cement kiln co-incinerators as a means to ensure environmental protection.

TABLE 13.2

Atmospheric emission limits — cement/lime kiln authorizations compared with a hazardous waste incinerator and the HWI Directive

Parameter	Directive 94/67/EC	Barrington (cement), AQ 1382
Particulate matter, mg m^{-3}	10	120
Oxides of nitrogen (as NO$_2$), mg m^{-3}	–	800
Oxides of sulphur (as SO$_2$), mg m^{-3}	50	1200
Carbon monoxide, mg m^{-3}	50	–
Hydrogen chloride, mg m^{-3}	10	10
Hydrogen fluoride, mg m^{-3}	1	1
Total organic carbon, mg m^{-3}	10	20
Dioxins and furans together, ng m^{-3}	0.1	1.0
Cadmium and thallium together, mg m^{-3}	0.05	0.1
Mercury, mg m^{-3}	0.05	0.1
Group III metals	0.5	1.0

Following trials of significant length at each of the cement kilns operating as co-incinerators, individual fuel specifications were inserted into the IPC authorizations for those kilns by means of a variation to the authorization. These specifications are reproduced in Table 13.3 on page 160. Cement kilns, with no scrubbing equipment, have a limited capacity to cope with chlorine, so the waste-derived fuel is limited in chlorine content. There are concerns about heavy metal levels in stack gases, in waste cement kiln dust and in the cement product, so heavy metal concentrations are restricted. In order not to add to the mass emissions of sulphur oxides, sulphur levels are also limited. The waste material is being used as fuel replacement so the calorific value, water and ash content are controlled. Entrained solids have different combustion characteristics, so entrained solid levels are kept down to a 30% maximum.

Ketton (cement), AS 3617	Ribblesdale (cement), AR 9435	Whitwell (dolomite), AR 0926	Cleanaway (incinerator), AG 8233***
50*	90/100	110**	20
1800/1650	1200	1750**	350
1350	(790) 2300	2500**	50
–	–	–	50
10	–	5**	10
2	–	–**	2
–	–	–**	20
1.0	1.0	1.0	1.0
0.1	0.1	0.1	0.1
0.1	0.1	0.1	0.1
1.0	1.0	1.0	1.0

* Limit applies from 1 July 1996
** Reduced limits or target figures to be introduced from 1 April 1997
*** Hazardous waste incinerator

ACTUAL PERFORMANCE OF INCINERATORS VERSUS CO-INCINERATORS

Up to this point, regulatory emission limits of incinerators and co-incinerators have been compared. The figures available from the public register on *actual* performance of two co-incinerators burning waste compared with a purpose-built incinerator are given in Table 13.4 on pages 162–163.

ENVIRONMENTAL EFFECTS OF THE USE OF CO-INCINERATORS

There are three main environmental effects of the use of cement kilns as co-incinerators.

159

TABLE 13.3
Summary of fuel specification*

	Barrington	Ketton	Ribblesdale	Whitwell
Max % replacement	25	20	40	25
Calorific value, mJ kg^{-1}	23–37	>21	23–29	21–27
Chlorine, % w/w	2	2.5	2	4
Sulphur, % w/w	0.5	0.5	0.3	0.5
Mercury, mg kg^{-1}	10	20	20	20
Cadmium and thallium, mg kg^{-1}	50	50	40	40
Heavy metals, mg kg^{-1}	1800	1800	1800	600
Fluorine, mg kg^{-1}	} 30,000	500	} 5000	3500
Bromine, mg kg^{-1}		300		4000
Iodine, mg kg^{-1}		100		100
Water, % v/v	Not detectable	1	Not detectable	–

* These are the key parameters; others may be included in the individual specifications

REDUCED USE OF FOSSIL FUELS IN CEMENT KILNS
The reduced use of fossil fuels represents an environmental benefit.

LESS RECOVERY OF USED SOLVENTS
The solvent recovery business in the UK has been seriously affected by this new, cheap disposal route for destruction of high calorific value liquid wastes. The cement kiln option means that material recycling and reuse is reduced and more virgin solvents are being purchased and used by industry. Virgin solvents require more raw materials — fossil fuels being one of those raw materials. This is obviously an environmental penalty.

POLLUTANT LOADINGS ON THE ENVIRONMENT
Certain pollutants are created by burning hazardous waste. Purpose-built incinerators, as explained earlier, are designed with the environmental fate of each

individual pollutant considered and catered for — for example, the vitrification of non-volatile pollutants and the efficient capture of particulates from flue gases. Because of the different regulatory regimes and process designs it is clear that for every tonne of waste burned there is potentially more pollutant load escaping into the environment from co-incinerators than from purpose-built incinerators. An example is dioxins, where calculations from figures on the public register show that the dioxin-specific design of Cleanaway's incinerator produces between 7 and 82 times less dioxins from each tonne of waste incinerated than the cement kilns. This is obviously an overall environmental penalty in terms of the best practicable environmental option (BPEO) for waste destruction, but the cement companies state that their *overall* emissions to the environment are no worse when burning a mixture of waste and coal than they would be burning coal alone. Depending whether the BPEO approach for waste destruction is taken, or the BPEO for the cement-making process, different answers are obtained.

THE CHOICE AHEAD

The UK and Europe seem to have currently positioned themselves reasonably with respect to the issue of co-incineration. In the USA, for various historical regulatory reasons and differences in principle, the cement industry now incinerates 70% of the hazardous waste produced, and purpose-built high temperature incinerators are closing down one by one due to financial non-viability. There are virtually no controls on the waste specifications for co-incineration in the US, and one cement works actually spent two years using their kilns primarily for waste disposal, feeding the minimum of raw material to the kilns and then landfilling the product due to its high toxicity.

At the other extreme, countries such as Finland, Holland and Switzerland only allow cement kilns to co-incinerate if they meet the same emission limits as purpose-built waste incinerators. This involves fitting expensive scrubbers and better particulate abatement plant. This is not usually financially viable — for example, a cement company in Holland estimated the cost of compliance at 300M guilders (about £120M) and the option was rejected. In Switzerland, the Holderbank cement company fitted the equipment, and now meets incinerator standards. The UK policy, allowing existing cement plants to burn only restricted waste mixtures which minimize potential environmental damage, appears a reasonable 'middle of the road' approach.

TABLE 13.4
Atmospheric emission comparisons for hazardous waste incineration plants and cement kilns

Parameter	Cement plant 1 25% waste, 75% coal
Particulates, mg m^{-3}	57–209
Total organic carbon, mg m^{-3}	1–5
Hydrogen chloride, mg m^{-3}	2–16
Hydrogen fluoride, mg m^{-3}	0.3–1.4
Oxides of sulphur (as SO$_2$), mg m^{-3}	50–7348
Oxides of nitrogen (as NO$_2$), mg m^{-3}	383–2345
Carbon monoxide, mg m^{-3}	125–1220
Cadmium and thallium, mg m^{-3}	0.003
Mercury, mg m^{-3}	0.002–0.006
Sb, As, Pb, Cr, Co, Mn, Ni, V, Sn, mg m^{-3}	0.14–0.2
Dioxin TEQ, ng m^{-3}	0.15–0.67

However, there remains a threat to the integrity of the UK's approach. Two of the UK cement companies currently burning waste have appealed to the Secretary of State against various conditions in their IPC authorizations. Both are requesting higher SO$_x$ emission limits and one is requesting more heavy metals to be allowed in the waste. The Secretary of State's decision on these appeals, due in late 1996, will be a reliable indicator as to whether the UK will move towards the US approach to co-incineration or maintain its environmentally justifiable middle ground.

Cement plant 2, 40% waste, 60% coal	Cleanaway's incinerator, 100% waste (1995 average)
42–109	2.06
29–39	< 2
26	0.26
0.15	< 0.1
883–1967	< 5
697–1784	177
326–984	10.2
No figures	Non-detectable
No figures	0.0023
1.3	Non-detectable
0.3	0.04

14. CONTROLLING VOC EMISSIONS USING LOW TEMPERATURE (CRYOGENIC) CONDENSATION

Craig Ibbetson

INTRODUCTION

The control of emissions to the environment is a significant issue in the process industries. Discharges from process plant can represent a threat to the environment, create problems with neighbours and waste valuable raw materials. Emissions of volatile organic compounds (VOCs) are under scrutiny because of the impact on ground level air quality, stratospheric effects and the effect on the local environment.

It can be difficult for the operator to select an abatement technique that is affordable and provides confidence that the required discharge standard will be met under a wide range of process conditions. The selection process is complicated due to the wide range of technologies to consider and process conditions encountered.

This chapter describes a technique for controlling VOC emissions from chemical process plant that uses low temperature condensation to achieve abatement and recovery of the widest range of organic compounds under varying plant conditions. The low temperature utility is provided by liquid nitrogen. If nitrogen is already used on site to provide an inert gas system, no additional nitrogen need be purchased to provide refrigeration and VOC control can be achieved with the lowest running costs of any technique.

This technology is already in use in a range of different process applications in Europe and North America.

WHY CONTROL VOC EMISSIONS?

The role of VOCs in ground level atmospheric pollution is clearly understood. Most industrialized economies are implementing control measures to limit VOC discharge to the atmosphere.

VOCs are a precursor in the formation of ground level ozone; an irritant at low concentration (WHO 8 hr TWA exposure 50 ppb). In order to form ozone two other components are required: NO_x and UV radiation. NO_x formation is related to the combustion of fossil fuels and is, in part, related to industrial activity such as power generation, petroleum refining and, most emotively, car use. UV radiation is supplied by the sun.

A substantial quantity of VOC emission is related to traffic movements, but — with a few noteworthy exceptions — the regulators of most countries have not yet devised a method of controlling emissions from these 'mobile sources'. Instead attention has been focused on large 'stationary sources' such as the coatings industry, fuel storage and distribution, and the chemical industry.

The reactions involved in the formation of ozone are simplified here:

$$\text{Sunlight} + NO_2 \rightarrow NO + O$$

The oxygen atom then combines with an oxygen molecule to give ozone:

$$O + O_2 \leftrightarrow O_3$$

Normally the ozone is short-lived because it reacts again with NO to return to the original starting point:

$$NO + O_3 \rightarrow O_2 + NO_2$$

VOCs interfere with this removal mechanism, however, since together with sunlight they create competing radicals which remove NO and thus prevent it from reacting with the ozone and returning it to oxygen molecules.

Organic compounds have differing abilities to form ozone and they are classified by their 'photochemical ozone creation potential' (POCP). Table 14.1 gives POCPs for a number of chemical classes.

Legislation in many countries reflects the POCP rating of different compounds together with other concerns such as toxicity and persistence in the environment.

LEGISLATION

The United Nations Economic Council of Europe has agreed a 30% reduction in VOC emissions by 1999 (from a 1988 basis). Each European country has incorporated this objective in different ways within its national laws. For example, Germany (TA Luft) and the UK (Environmental Protection Act 1990) have adopted discharge limits based on the industrial process concerned. Other countries have adopted similar standards.

The USA has adopted regional or state controls driven by the local or regional air quality. Industry located in an 'ozone non-attainment area' is set

TABLE 14.1
Photochemical ozone creation potential (POCP) for a number of chemical classes

Compound	POCP
Alkenes	84
Aromatics	76
Aldehydes	44
Alkanes	42
Ketones	41
Esters	22
Alcohols	20

significantly higher standards than another in an area of higher air quality. Most concern in the USA is in the North East, Mid West and California.

It is worth noting that different countries have different approaches to setting numerical standards. In the US there is a clear preference for standards based on mass — that is, each discharge must be reduced by a given amount (typically 95% or 98%). The regulations employed in the UK and Germany drive towards limits based on concentration but may employ mass reductions as 'mileposts' along the way. Normally the use of concentration-based standards implies higher standards for industry and the suppliers of abatement technologies. Note that for chemical manufacturers the application of standards based on concentration may not allow the use of technologies that have been developed for markets where mass-based standards are employed, due to the difference in reduction efficiency that is implied when outlet concentrations are specified by the regulator. Instead of having to achieve 95% to 98% reduction efficiency, these standards may require 99.9% reduction or greater. A USEPA[1] document outlines the control efficiencies typically achieved with a range of technologies (see Figure 14.1 on page 168). Cryogenic condensation has been added for comparison.

Most countries embrace a hierarchy of waste control that encourages operators of processes to consider first waste control measures followed by recovery/reuse and then destruction as the last option. Table 14.2 on page 168 shows the applicability of the alternative abatement techniques. Following this procedure has the benefit of minimizing the cost of implementation of the abatement unit.

CONDENSATION

The use of condensation to control vapours is well known and widely practised throughout the process industries. The low temperature utility is typically chilled water, brine or glycol, or occasionally a low temperature heat transfer

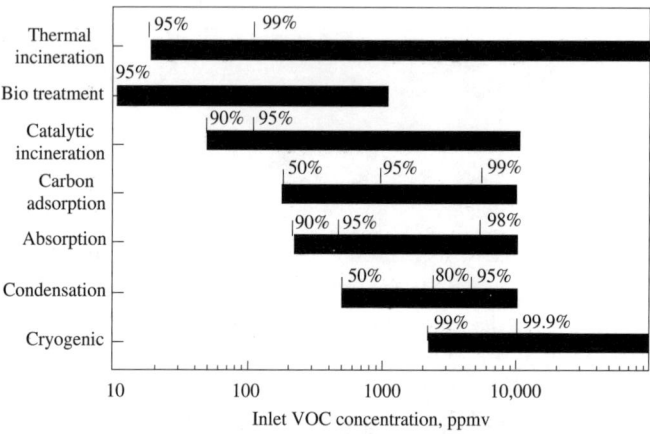

Figure 14.1   Reduction efficiencies of different abatement techniques.

TABLE 14.2
Applicability of the alternative abatement techniques

	Condensation	Absorption	Adsorption	Oxidizer
Elimination				
Reduction	✓			
Reuse	✓			
Recovery	✓	✓	✓	
Destruction				✓

TABLE 14.3
Equilibrium temperatures for a range of VOCs

Compound	Equilibrium temperature for UK guideline standards, °C
Acetone	-97
Methanol	-74
Methylene chloride	-118
Toluene	-72
Methyl iso-butyl ketone	-75
Tetra hydro furan	-93

fluid. The cost of these utilities is relatively high and is limited by the evaporator temperature of the primary refrigerant. The cost of low temperature refrigeration plant rises rapidly when temperatures lower than $-40°C$ are required, as the complexity of the compression equipment increases.

When the objective is to control compounds with high vapour pressures then it is unlikely that a conventional refrigeration system will be able to reduce the outlet concentration to a value low enough for environmental compliance. As illustration the equilibrium temperatures for a number of VOCs are shown in Table 14.3. It can be seen that it is not economically justifiable to use a conventional refrigeration system to reach the required temperatures — a lower temperature utility is required. This has traditionally been considered to be prohibitively expensive, but ignores a potential source of low temperature which already exists in the nitrogen storage vessel.

LIQUID NITROGEN

Liquid nitrogen is produced by the separation and liquefaction of the components of air, and is produced by distillation at low temperature followed by liquefaction using a compression and expansion cycle. Liquefaction results in a near 700-fold reduction in volume, reducing the cost of transportation from the air separation unit to the user.

At the point of use the nitrogen is vaporized in a heat exchanger to convert it back to the gaseous state. The energy required to do this is supplied

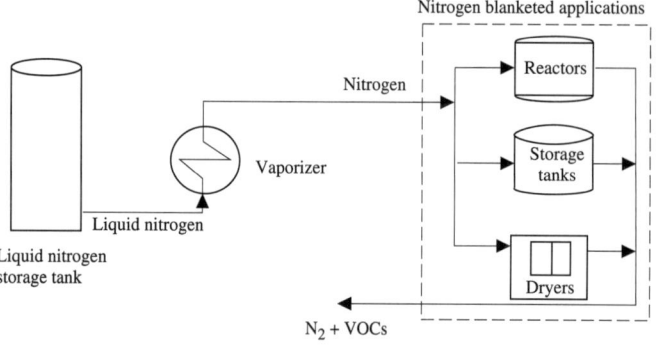

Figure 14.2   Schematic of liquid nitrogen supply.

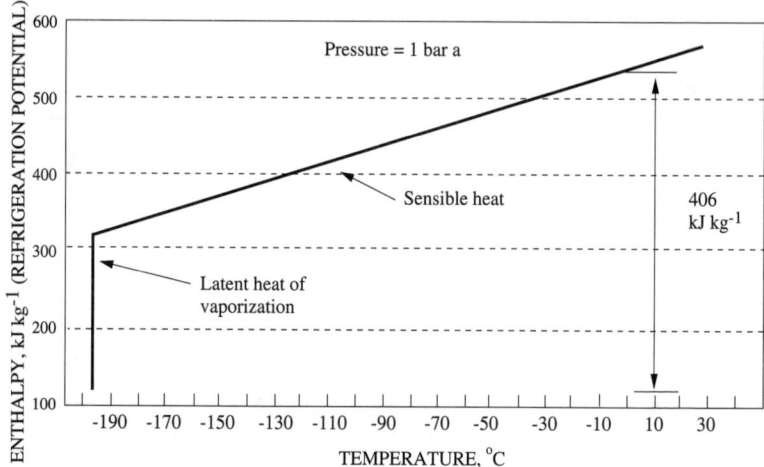

Figure 14.3   Enthalpy/temperature diagram for liquid nitrogen.

Example: saturated liquid at $-196°C$ when vaporized to gas at $0°C$ and atmospheric pressure will provide $(531 - 125)$ kJ kg^{-1} of refrigeration potential $= 406$ kJ kg^{-1}.

by the ambient air or from low pressure steam if a higher duty is required. Figure 14.2 shows a schematic of nitrogen supply.

A substantial fraction of the cost of producing liquid nitrogen is associated with the liquefaction process. At atmospheric pressure the latent heat of vaporization is 199 kJ kg^{-1}, the sensible heat is approximately 1 kJ kg^{-1} C^{o-1} and the boiling point is $-196°C$. See Figure 14.3.

In the past it was not easy to use economically the heat transfer associated with vaporizing nitrogen because the cost of the heat exchange system was high. This inhibited use of liquid nitrogen, except in a few installations in the fine chemicals industry where it is used to provide cooling of reaction vessels to low temperatures.

With the advent of controls on VOC emissions the high capital cost of all alternative abatement techniques makes it viable to consider using low temperature condensation based on liquid nitrogen, as it offers substantial savings in operating cost and the ability to recover and reuse the condensed product.

WHEN TO USE LOW TEMPERATURE CONDENSATION

The most important factor governing the applicability of this technique is the availability of sufficient liquid nitrogen to provide refrigeration for the abatement unit. In the process industries liquid nitrogen is widely used as the sole source of inert gas or as a back-up to on-site gas generators.

From economic considerations this technique should be considered when the contaminated air flow is in the range 0 to 5000 Nm^3 hr^{-1}. Concentrations of VOCs can range from zero to fully saturated; composition is of secondary importance provided the appropriate control temperature is known. Air flow can be minimized by the use of nitrogen to replace air as the process carrier gas, resulting in reduced capital and operating costs through higher VOC concentrations, as the absence of oxygen avoids the need to dilute below 25% of the Lower Explosive Limit (LEL)[2]. In most cases the energy associated with vaporization of the nitrogen carrier gas provides sufficient energy to condense the resulting VOCs (see Figure 14.4).

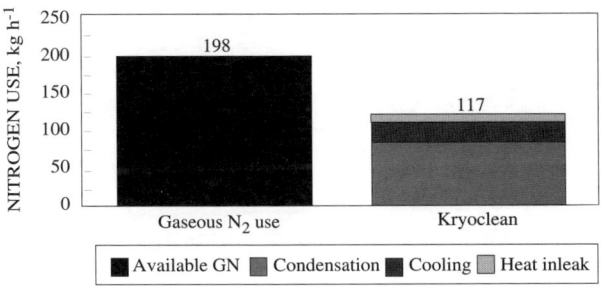

Figure 14.4   Typical comparison of liquid nitrogen (LN) demand for condensation against site gaseous nitrogen (GN) use.

The advantage of the low temperature condensation technique is that it can deal with all organics (even in the presence of water), produces no secondary waste streams and can function when the concentration and composition are changing over time. This feature makes it particularly suitable for VOC control in the fine chemicals industry where multi-product, multi-purpose plants are widely used and batch processing is the norm. Other applications are in product loading, and in some continuous processes where solvents are used.

ENGINEERING ISSUES

The design of condensing systems is well understood, but the correct specification of a low temperature condenser is not driven by heat transfer considerations alone. Low temperatures result in a gradual accumulation of frozen material on the heat exchanger surface. Also, the potentially large temperature driving forces can create 'mist' which can be carried out of the condenser and cause it to fail to comply with the discharge standard. The design of the condenser is crucial to the successful operation of the abatement unit and represents the proprietary knowledge of the equipment vendor.

A successful condenser design has the following features:

• careful control of exchanger surface temperatures to minimize freezing;
• accurate control of temperature differences to minimize mist formation;
• fast response to process changes in order to remain in compliance (whilst maintaining control of the two points above);
• a defrosting system which removes frozen material from the heat exchange surface and recaptures it.

BOC GASES KRYOCLEAN — SYSTEM OPERATION

Process flow

Figure 14.5 shows the process flow being cooled by a recirculating flow of cold nitrogen gas in the working condenser (exchanger HE–200 or HE–300). At the outlet of HE–200/300 the process stream will have reached the desired operating temperature and the condensate will have drained by gravity out of the bottom of the exchanger. The cold vapour stream is rejected to vent after an economizer has recovered the cold by exchanging it with the coolant recirculation flow.

Coolant flow

Liquid nitrogen is used to provide cooling and motive force for the gaseous

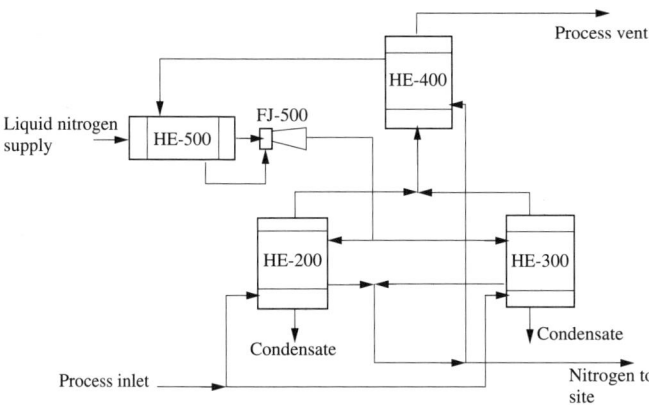

Figure 14.5 Kryoclean process flow diagram.

nitrogen circulation system. The vaporized nitrogen does not contact the process vapour and is available for reuse on site as pure, inert, high pressure gas.

Liquid nitrogen is vaporized in HE–500 and is mixed with the recirculation flow in ejector FJ–500. The combined flow is admitted to condenser HE–200/300 as a cold gas. At the exit of HE–200/300 the temperature has increased and the nitrogen flow divides in order to allow more cold gaseous nitrogen to be admitted. The vent stream is available for plant use; the recirculation flow is cooled by the cold exhaust from the process and returned to HE–500 in order to vaporize the incoming liquid nitrogen.

The principle of using gaseous nitrogen as the cold working fluid is to avoid the very low temperatures encountered with boiling liquid nitrogen. The design of the system recovers the energy available from the vaporizing liquid nitrogen in order to ensure the most efficient use of energy. By using this design it is possible to control the tube wall temperature in the condenser more closely and this greatly assists in the mechanism of condensation and avoids localized freezing. The temperature of the recirculating stream can be set to the desired temperature by control of the liquid nitrogen input rate, which in turn controls the process vent exhaust temperature.

After a number of hours of operation an amount of frozen material will have accumulated on the surface of the heat exchanger and will have fouled it beyond its design rating. (Only some of the 'freezable' material that enters the condenser freezes on the heat exchange surface. Most material condenses as a liquid and runs out of the exchanger, provided that tube wall temperatures are controlled properly.) At this point the condensers reverse; the on-line is taken out of service and is thawed by a heated slip stream of clean nitrogen. The off-line unit will have been pre-cooled to the correct operating temperature ready to come on-line when required.

Engineering

Low temperature condensation systems are compact pieces of equipment which are usually fabricated off site and brought onto site as a skid-mounted module. Engineering materials are usually low temperature grade stainless steel (for example, 316L) or an alloy such as monel if higher corrosion rates are anticipated. In the field the low temperatures mean that corrosion rates are low — BOC Gases has examined the cores of some its units after two years exposed to chlorinated species without finding any evidence of corrosion.

Utilities requirements are for liquid nitrogen, gaseous nitrogen to site, electrical power and steam for thawing.

CONCLUSIONS

Low temperature condensing systems based on liquid nitrogen represent an effective and proven method for minimizing VOC emissions from chemical processes. For plants which already use nitrogen it is the technique with the lowest operating cost. The flexibility of condensing systems means that they are particularly suitable for applications where the stream conditions are not constant due to changes in flow, concentration or composition. By changing the operating temperature it is possible to reduce outlet concentration, thus enabling future compliance with tighter standards than are required today.

REFERENCES IN CHAPTER 14
1.   USEPA, June 1991, *Handbook of Control Technologies for Hazardous Air Pollutants, EPA/625/6–91–014* (USEPA, Cincinatti, OH, USA).
2.   NFPA, 1992, *Standard on Explosion Prevention Systems, Report no 69* (NFPA, Quincy, MA, USA).

15.   VOC DESTRUCTION IN A MICROWAVE-ENERGIZED PLASMA

Jim Hutchison and Neil Wright

INTRODUCTION

Organic solvents are extensively used in industry, across a wide variety of sectors. Many processes hinge on their use, but this leads to the production of large quantities of effluent laden with volatile organic compounds (VOCs), whose impact on the atmosphere is now of worldwide concern. This chapter considers the application of a microwave-energized plasma as a novel technique for the destruction of VOCs.

ENVIRONMENTAL EFFECTS OF VOCs

There are at least four potentially damaging effects which arise from the discharge of VOCs to the atmosphere.

PHOTOCHEMICAL OZONE CREATION IN THE LOWER ATMOSPHERE (TROPOSPHERE)

Ozone concentrations near the ground can be raised to levels considered dangerous to human health (especially to those with respiratory problems) and damaging to vegetation, by a complex series of chemical reactions involving the interaction between VOCs, oxides of nitrogen (NO_x — for example, from motor vehicle exhausts) and sunlight.

$$VOCs + NO_x + sunlight \rightarrow ozone + other\ photochemical\ pollutants$$

Examples of other photochemical pollutants are formaldehyde, acetic acid and peroxyacetyl nitrate (PAN).

CONTRIBUTIONS TO GLOBAL WARMING

The more stable VOCs, which do not undergo oxidation in the atmospheric boundary layer (0.5–2 km), are transported into the free troposphere above the boundary layer, where they may be accumulating. If any of these compounds absorb solar or terrestrial infrared radiation, they would then be classed as radiatively active gases and may contribute significantly to global warming. For example, the global warming potential (GWP) of perfluoroethane = 12,500 times the GWP for carbon dioxide (CO_2).

VOCs which are not themselves classed as radiatively active gases may perturb the concentrations of gases which are radiatively active — for example, by taking part in reactions which produce ozone (a 'greenhouse' gas) in the free troposphere.

DEPLETION OF OZONE IN THE UPPER ATMOSPHERE

Depletion of ozone in the upper atmosphere (stratosphere) allows higher levels of harmful ultraviolet radiation to reach the earth's surface.

Organic compounds stable enough to survive removal processes in the troposphere (~0–15 km) eventually reach the stratosphere (~15–50 km). If these compounds contain chlorine or bromine atoms, their photolytic decomposition in the stratosphere leads to the formation of ozone-destroying 'chain carriers':

$$Cl\cdot + O_3 \rightarrow ClO\cdot + O_2$$

$$ClO\cdot + O \rightarrow Cl\cdot + O_2$$

The production and supply of a number of once commonly used solvents, refrigerants and propellants, now known to have high ozone depletion potentials, have been or are being phased out under the Montreal Protocol and subsequent international agreements. These substances include chlorofluorocarbons (CFCs), Halons (for example, methyl bromide) and 1,1,1-trichloroethane.

DIRECT EFFECTS ON HUMAN HEALTH

Compounds known to be 'air toxics' and to be widely distributed in the ambient atmosphere include benzene, 1,3-butadiene (both potential leukaemia-inducing agents), formaldehyde (potential nasal carcinogen), polynuclear aromatic hydrocarbons (PAHs, potential lung cancer-inducing agents), polychlorinated biphenyls (PCBs), dioxins and furans.

LEGISLATION

Research continues on the nature and extent of the environmental problems caused by VOCs. A number of legislative and other measures impact on the control of VOC emissions in the workplace and the environment. In the UK,

these include the following:
- The Environmental Protection Act 1990 (EPA 90). This prescribes VOCs for release to air. While the act sets no emission levels, these are found in guidance from either the Chief Inspector of Pollution (for Part A processes), or from the Secretary of State (for Part B processes). While the emission levels given in Part A guidance are not mandatory, those for the Part B processes are. The levels set vary from sector to sector, but would typically be a concentration of about 50 mg m^{-3} of VOC expressed as carbon;
- The Control of Substances Hazardous to Health (COSHH) Regulations. These control workplace exposure to various materials, with allowable exposure levels set by the HSE;
- the UK is committed under the VOC protocol of the 1991 Long Range Transboundary Air Pollution (LRTAP) Convention to a 36% reduction in VOC emissions by 1999.

VOC ABATEMENT TECHNOLOGIES
Coatings industries can adopt either compliance coatings or compliance equipment to meet legislative requirements. The main candidates for the first option are water-borne and powder coatings. Compliance equipment is normally classified into the following five categories:

THERMAL INCINERATION
Support fuel is used to heat the VOC-contaminated stream in air to between 760°C and 870°C, producing carbon monoxide, water and heat, which is usually recovered to reduce costs. Halogenated solvents require higher temperatures and further exhaust treatment. Conversion efficiency is nearly 100%. The high thermal mass of the refractory linings makes the technique better suited to continuous processes.

CATALYTIC INCINERATION
Catalytic incineration is similar to thermal incineration except that catalysts reduce the temperature to between 250°C and 420°C, obviating the need for refractory linings. The method cannot cope with large quantities of halogenated VOCs.

ADSORPTION
The exhaust stream is passed over a porous granular adsorbent solid with a large

surface to volume ratio, usually activated carbon, and the VOCs are rapidly extracted onto the surface of the solid. The adsorbent can be regenerated by heating, either with steam or hot air, and the VOCs condensed and recovered, possibly for reuse. In the adsorption process, prediction of removal efficiency is sometimes difficult.

ABSORPTION

The VOCs are removed in a scrubbing tower by dissolution in a liquid which runs over a bed of packing material contained in the tower, giving a high contact area. The liquid is then either disposed of after further treatment, or cleaned and reused.

CONDENSATION

VOCs are condensed at low temperature either on a heat exchanger surface or on the surface of sprayed droplets. The condensed solvent can sometimes be reused, but for multicomponent VOC emissions this is often not practical.

PLASMA TECHNOLOGY

Plasma technology is developing rapidly across a wide variety of industrial fields, including engineering, effluent treatment, and semiconductor production. As an alternative means of providing energy for industrial processes, the inherent activating nature of plasmas gives them unique capabilities in promoting gas phase reactions. Essentially, a plasma is an excited gas in which molecules have become dissociated, producing a mixture of ions, electrons and neutral species in various combinations.

The nature of the plasma can vary enormously depending on the conditions and the way in which these components are produced. The necessary electrical energy can be provided by direct current (DC), radio frequency or microwave sources, but the nature of the plasma, which dictates its field of application, is determined mainly by the pressure. Conventionally, plasmas are classified into one of two types — thermal or non-thermal. The latter is also frequently referred to as 'cool', 'cold', or 'non-equilibrium'.

THERMAL PLASMAS

At 'high' pressure, which in plasma terms can be taken as an absolute pressure of 10 mm Hg and above, high electric field intensity is required to create the breakdown, resulting in extensive ionization. The comparatively high pressure causes a high collision rate, resulting in substantial energy exchange between particles. The system therefore approaches thermal equilibrium and will have a very high temperature: from several hundred to more than 10,000°C.

Thermal plasmas are used in applications where high inputs of thermal energy are required. They are typically high pressure, high temperature devices, generated between high voltage electrodes or in high inductive fields, and normally delivering large amounts of power. Typical applications include thermal spraying, waste destruction, welding and arc sources. Usually these are atmospheric pressure processes, such as metal cutting, welding, thermal spraying of ceramics, and waste destruction.

NON-THERMAL PLASMAS

At low pressure, the mean free path is many orders of magnitude greater, and the collision rate and energy exchange correspondingly much lower. This allows the field to accelerate charged particles to velocities which depend on their mass, creating a distribution of energies which departs markedly from thermal equilibrium. Because of their greater mobility, lighter components (and in particular electrons) may attain kinetic energies which are much higher than if they were in thermal equilibrium with the bulk of the plasma.

The energetic electrons and their collision products activate unique reactions which thermal plasmas are incapable of promoting. As part of a comparatively cool bulk medium, they can lower reaction temperatures and facilitate processing at comparatively low thermal energies. Unfortunately, the vacuum operation normally required to generate non-equilibrium plasmas increases costs in two ways. Firstly the equipment is inherently expensive; and secondly, vacuum operation necessitates inefficient batch processing.

THE ATMOSPHERIC PRESSURE NON-EQUILIBRIUM PLASMA (APNEP)

EA Technology has discovered a new method of plasma production which is capable of generating non-equilibrium conditions at atmospheric pressure, thereby removing the disadvantages of vacuum operation. It achieves this highly desirable combination using cheap, simple, microwave equipment, giving it further economic benefits. A theoretical assessment based on some fundamental observations has confirmed its predominantly non-equilibrium nature, and initial work has demonstrated its ability to promote processes typical of non-equilibrium plasmas.

The basic features of APNEP are:

- a multi-mode microwave cavity adapted from a commercial oven;
- a power source adapted from commercial microwave components;
- specially designed plasma containment;
- plasma initiation equipment.

Electrons and ions generated by initiation equipment absorb the microwave energy and rapidly develop into a diffuse glowing plasma, which is sustained on cessation of the initiation process. The glow indicates continual recombination and relaxation of excited species, which are simultaneously replenished by the microwave field. Modifications to the proprietary microwave oven are not major, and the widespread availability of industrial microwave components means that the device will retain its low cost even after development for industrial use.

This plasma is unusual in that it does not appear to fall wholly into either of the categories of plasma consistent with conventional experience. It is sustained continuously at atmospheric pressure — typical circumstances for generation of a thermal plasma — yet it operates at much lower temperatures and has the appearance of a glow discharge normally associated with non-thermal plasmas. These factors suggest that it has the activating properties required for non-equilibrium surface processing, while its atmospheric operation renders it more amenable to continuous processing than competing vacuum plasmas.

In normal atmosphere the plasma glows quite strongly in the visible region, but appears to be separated from its containment vessel by a less bright sheath region several millimetres thick. This is a characteristic of glow discharges in which a substrate or wall in contact with the plasma tends to be at a lower potential than the plasma itself. Positive ions move preferentially towards this lower potential, while electrons are repelled, disturbing the macroscopic electrical neutrality in the space near the substrate. Only a few highly excited electrons will be able to penetrate this region, and since the plasma glow is caused by relaxation of atoms excited by electron impact, the comparative paucity of such events in the sheath region gives it a darker appearance.

From calculations based on simple observations it is estimated that the temperature of the atoms and ions at the centre is likely to be a few thousand °C, falling to several hundred °C at the perimeter. Absorption of the incoming energy is of the order of 40%. The degree of ionization is low, and it is difficult to explain the observed phenomena by thermal mechanisms alone, lending support to the non-equilibrium view.

The development of APNEP is underpinned by a fundamental investigation which is part of the Post Graduate Training Partnership run by EA Technology and UMIST with support from the DTI. Conclusions so far are as follows:

- the power primarily influences the volume of the plasma;
- the rate of decay of optical emissions depends on pressure;
- the main energy losses from the system are by diffusion to the walls and recombination of the charged species;
- the plasma can extend more than ten centimetres beyond the microwave cavity;
- microwaves are absorbed mainly in the surface of the plasma because of its high density;
- the plasma is predominantly non-equilibrium.

Future work will identify the main species and determine more precisely the ion density and electron energy.

The unusual combination of non-equilibrium conditions and atmospheric operation, as offered by APNEP, is crucial in enabling industry to exploit widely the benefits of plasma processing. Some specialized methods do exist, such as corona discharges for pretreatment of packaging film. Other techniques under research include pulsed RF plasmas, surface wave plasmas, annular slotted waveguide applicators, focused microwave pulses and lasers in a resonant cavity. While all these approaches are of value, it is felt that APNEP, through its comparative simplicity and flexibility, offers excellent prospects for industrial implementation.

VOC DESTRUCTION EXPERIMENTS

Figure 15.1 (see page 182) shows the arrangement used for carrying out the VOC destruction tests. The plasma was created in a quartz glass vessel (volume ~ 1 litre) contained inside a modified commercial microwave oven (power input 1.2 kW). Gas streams were introduced to the top of the vessel through an inlet glass tube which contained a baffle arrangement designed to encourage the gas to flow around the periphery of the plasma. This helped to avoid the plasma becoming destabilized at high gas flow rates. The treated gas stream from a glass outlet tube sealed into the bottom of the vessel was passed into a 5 litre flanged container which was quickly covered with a lid at the end of the sampling period. This design has by no means been finalized. Alternative configurations are being investigated and improvements to the sampling procedure are also in hand.

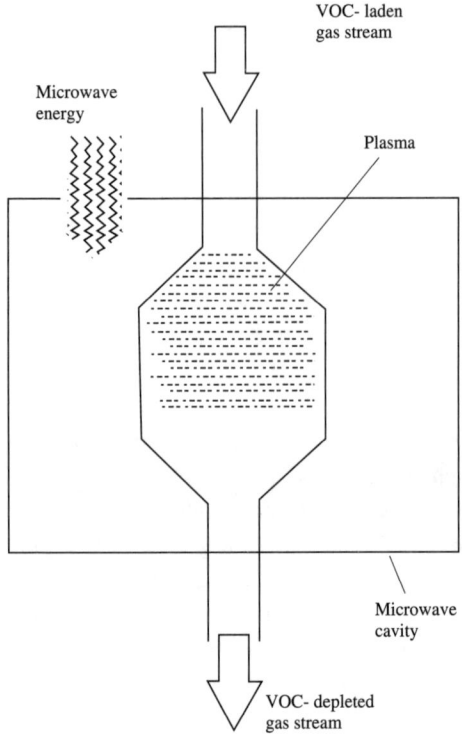

Figure 15.1   Schematic view of VOC removal in the microwave-energized plasma 'APNEP'.

The desired concentrations of each of the solvents tested were obtained by injecting calculated volumes of solvent vapour into the air stream which was being pumped into a 40 litre capacity 'Tedlar' (fluoropolymer) gas bag. The latter was held within a sealed rigid outer container. By blowing compressed air into the rigid container at the required flow rate, the flexible gas bag was made to collapse and the contents driven out to the plasma unit at the same flow rate.

The concentrations of solvent in the gas streams before and after passing through the microwave plasma were measured using a Research Engineers Limited 'Gas-Tec' flame ionization detector (FID). This could be operated using two concentration scales calibrated as 0–1000 ppm and 0–10,000 ppm methane. Because the scales were non-linear, with low concentrations having disproportionately large divisions, readings down to approximately 0.1 ppm could be discerned on the more sensitive range. The variation of the FID reading with solvent concentration was checked and found to be broadly linear.

For early experiments in which low flow rates ($1–2$ l min^{-1}) of air containing acetone, toluene and trichloroethylene at concentrations covering the range 200–1000 ppm by volume were passed through the plasma, the removal of solvent achieved was at least 99% in all cases (see Table 15.1). When air containing 1000 ppm by volume of butanone (methylethylketone, MEK) was passed through the plasma at the much higher flow rate of 1 m^3 hr^{-1} (16.7 l min^{-1}), the removal of the butanone was still 95%.

Following further refinements to the equipment, a more comprehensive series of tests was carried out using the hydrofluorocarbon 1,1,1,2-tetrafluoroethane (TFE). Because of the number of highly stable C–F bonds present in this compound, it was expected to be more resistant to oxidation than the more commonly used solvents, and so provide a sterner test for this technology. A range of concentrations and flow rates were investigated (Table 15.2, page 184). The plasma was successful in bringing about essentially complete destruction of the TFE at gas flow rates of 1 m^3 hr^{-1} and 2 m^3 hr^{-1} and inlet concentrations up to 15 g m^{-3} (the highest tested). At 5 m^3 hr^{-1} flow rate, destruction of the TFE was incomplete, but still of the order of 90% or more, for all three concentrations tested. For the 15 g m^{-3} concentration, the threshold flow rate above which breakthrough of the TFE occurred was found to be approximately 3 m^3 hr^{-1}.

TABLE 15.1
Treatment of various solvent-air mixtures in microwave-energized plasma

Solvent	Concentration of solvent in air		Flow rate of gas stream through plasma		Removal of solvent*
	mg m^{-3}	ppm by vol	m^3 h^{-1}	l min^{-1}	
Acetone	480	200	0.06	2.0	> 99%
Toluene	3840	1000	0.03	1.0	> 99%
Trichloroethylene	3960	720	0.06	2.0	> 99%
Butanone (MEK)	3000	1000	1.0	16.7	> 95%

* (from FID readings before and after treatment)

TABLE 15.2
Treatment of 1,1,1,2-tetrafluoroethane at various concentrations and flow rates

1,1,1,2-TFE concentration		Flow rate of gas stream through plasma		Removal of 1,1,1,2-TFE*
mg m^{-3}	ppm by vol	m^3 h^{-1}	l min^{-1}	
500	120	1.0	16.7	> 99.8%
500	120	2.0	33.3	> 99.8%
500	120	5.0	83.3	99.2%
1500	350	1.0	16.7	> 99.95%
1500	350	2.0	33.3	> 99.95%
1500	350	5.0	83.3	96.5%
15,000	3500	1.0	16.7	> 99.99%
15,000	3500	2.0	33.3	> 99.99%
15,000	3500	3.0	50.0	99.8%
15,000	3500	4.0	66.7	97.6%
15,000	3500	5.0	83.3	93.6%

* (from FID readings before and after treatment)

CONCLUSIONS
The potential of the microwave-energized plasma ('APNEP') for destroying VOCs has been clearly demonstrated for a variety of concentrations of different solvents introduced to the plasma at flow rates (so far) up to 5 m^3 hr^{-1}.

APNEP is at an early stage of development, but is expected to progress rapidly towards industrial applications. In developing the engineering for efficiency, reliability, flexibility and control, the following issues will be investigated:

• optimization of cavity size and shape;
• modifications to the microwave source to maintain stability at all power levels;
• increasing the non-equilibrium proportion of the plasma — for example, by vessel design and flow control, and further electrical modifications;
• determination of the relationships between flow rate, power, and plasma shape;

- development of the plasma initiator system, for control and flexibility;
- durability engineering of the system;
- development of relevant monitoring techniques. These may include spectroscopic methods, or inferred parameters from simpler measurements as appropriate.

This further development of the technology is being undertaken within a multi-client programme involving equipment manufacturers and companies from the printing, surface coating and other solvent-using sectors urgently seeking solutions to their VOC emission problems.

ACKNOWLEDGEMENTS
The authors wish to acknowledge the contribution of their colleague Dr X. Duan, who discovered this form of plasma. Thanks are also due to F.M. Taylor, A.J. Lacey and M.H. Littlewood for constructing the equipment and carrying out the experiments, and to W.B.R. Moore and M.I. Colley for their helpful discussions.

16. REDUCTION OF CHLORINATED ORGANIC COMPOUND EMISSIONS BY FLAMELESS THERMAL OXIDATION*

John Young

INTRODUCTION

Thermatrix technology is a unique, proprietary, patented technology for the flameless thermal oxidation (FTO) of noxious emissions which arise in the normal course of operations in the oil and gas, chemical, pharmaceutical and manufacturing industries. Thermatrix applies its thermal oxidation technology for the highly efficient, controlled, non-flame oxidation of chlorinated volatile organic compounds (CVOCs) in a ceramic matrix called a 'packed bed'[1]. The oxidation of organics occurs in a 'reaction zone' contained within the bed of chemically inert ceramic materials typically operated at 850–950°C.

The company has developed several flameless oxidizer styles to meet specific treatment needs. These oxidizer styles include electrically-heated units generally used in treating CVOC streams with flow rates up to 250 m^3 hr^{-1}, and gas-heated, straight-through and recuperative styles typically used in the treatment of CVOC streams with flow rates greater than 250 m^3 hr^{-1}.

The gas straight-through style oxidizer (see Figure 16.1 on page 188), consists of an insulated metal shell containing a heated ceramic matrix. In operation, the CVOC stream and any air required to support the oxidation reaction passes into the bottom of the preheated bed and moves upward through the matrix. The temperature of the incoming gas rises as it picks up heat from the bed until the oxidation temperature of the organic compounds is attained. Once the reaction temperature has been reached, the organics in the CVOC stream oxidize creating a stabilized reaction zone as heat is given up to the surrounding matrix.

The large thermal mass of the bed also enables it to store or release large amounts of heat without rapid changes in temperature. In many cases the CVOC stream may already contain sufficient heating value to sustain the bed

---

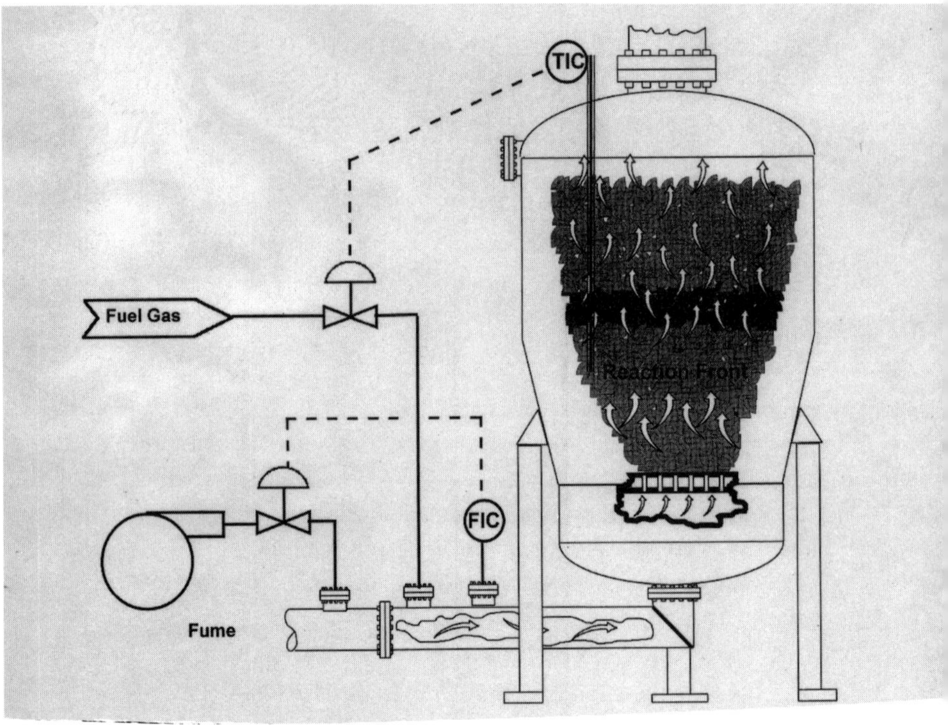

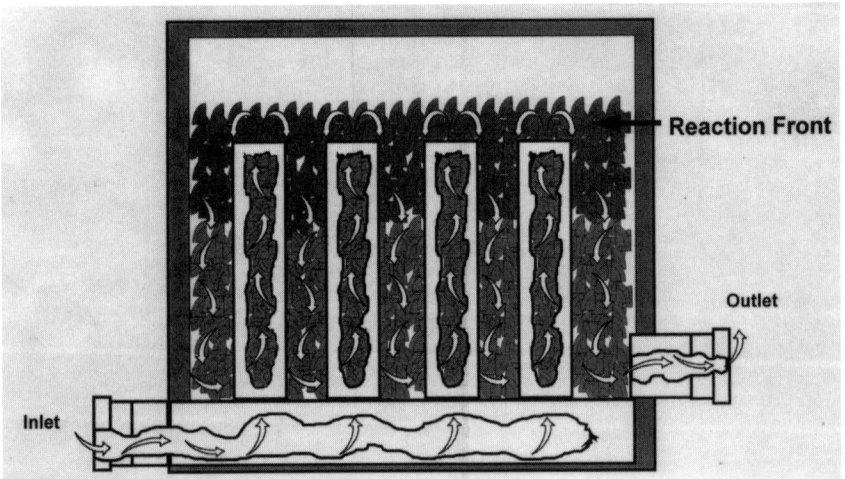

Figure 16.1 Simplified elements of process control.

Figure 16.2 Flameless thermal oxidizer with internal heat recovery.

temperatures. If needed, supplementary energy can be provided from either an electrical heater or by enriching the mixture with natural gas or propane.

Figure 16.2 schematically presents a basic technology enhancement — that is, internal oxidation heat recuperation. Heat recuperation in a Thermatrix thermal oxidation unit is accomplished by flowing the incoming and outgoing gases counter-currently through the bed with metal tube separation[2]. In this manner, heat produced during oxidation of the organic compounds is used to raise the temperature of the incoming gas mixture. The exit gas temperature of the recuperative style reactor is typically 300–400°C — significantly lower than 850–950°C gas straight-through oxidizer exit gas temperature.

The FTO technology has been integrated in systems designed to destroy VOCs effectively and remove any environmentally offensive off-gases formed on oxidation. Typical integrated system components include the inlet gas train (wherein the CVOC stream, air and auxiliary fuel, if needed, are premixed prior to treatment), the oxidizer, a quench and scrubber (if required) and a stack.

Thermatrix has designed, built, installed and commissioned systems based on CVOC treatment by flameless thermal oxidation. One system is treating CVOCs extracted from a soil vapour extraction (SVE) well in a remediation project. Another is currently being used to treat the CVOC gaseous effluent from a waste water air stripping process. Two large FTO systems are being used to treat process vent streams containing volatile chlorinated hydrocarbons from herbicide manufacturing processes.

PERFORMANCE OF FTO IN TREATING CHLORINATED VOCs

FTO is a very effective demonstrated technology for the treatment of fume streams containing CVOCs. Extensive FTO performance testing on CVOCs was conducted during 1995[3]. The testing was done with an ES 300H (electrically heated) unit. Figure 16.3 (page 190) presents a schematic of the test equipment configuration. The equipment configuration enabled treatment of CVOCs extracted from effluent from a contaminated well head and also included a provision for spiking the feed stream with trichloroethylene (TCE), tetrachloroethylene (PCE) and 1,1,1-trichloroethane (TCA).

In 22 days of continuous operation treating the well head effluent, a total of 11.17 kg of total CVOC was destroyed with no identifiable products of incomplete oxidation observed in any outlet sample at a concentration of greater than 2 ppbv — that is, the detection limit. In 2.5 days of tests in which the feed

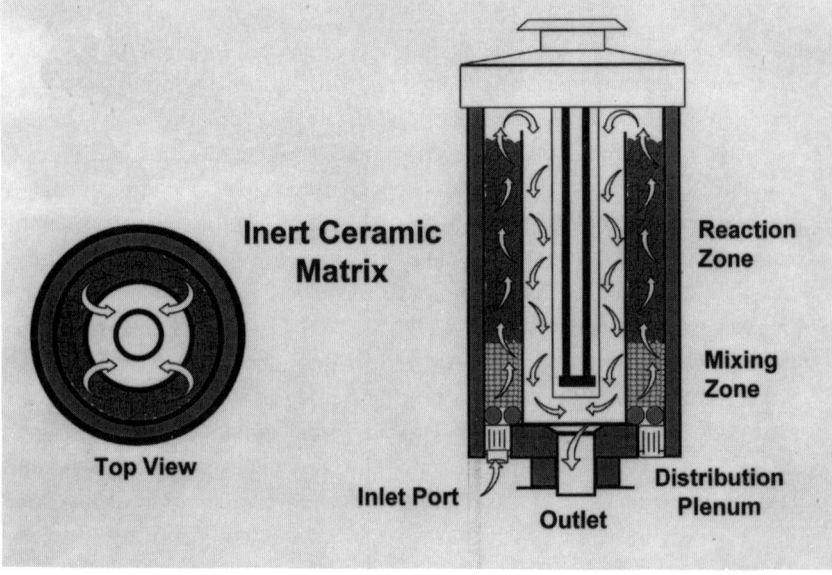

Figure 16.3   Flameless thermal oxidizer (with electric preheat).

TABLE 16.1

Flameless thermal oxidizer performance — results of CVOC spiking tests

	PCE, ppmv	TCE, ppmv	TCA, ppmv	Total CVOC[b], ppmv
Spike 1	551	279	204	1126
Spike 1 DRE[a]	> 0.999982	> 0.999856	> 0.999804	> 0.999913
Spike 2	1037	607	456	2182
Spike 2 DRE[a]	> 0.999990	> 0.999934	> 0.999912	> 0.999957
Spike 3	1915	778	386	3087
Spike 3 DRE[a]	> 0.999995	> 0.999949	> 0.999896	> 0.999971

[a] outlet analysis for all primary constituents were non-detect. Reported destruction and removal efficiencies (DREs) are minimum values (that is, DRE > listed value)
[b] total CVOC primary constituents listed in table

stream was spiked with PCE, TCE and TCA combined concentrations of up to 3080 ppmv, over 1.5 kg of total CVOC was destroyed. The combined CVOC destruction and removal efficiency measured in these tests was greater than 99.99%. Table 16.1 presents the results of the spike tests.

FTO SYSTEM PROJECT CASE HISTORIES

TREATMENT OF WASTE WATER CONTAINING CVOCs

In an effort to voluntarily reduce emissions, a major chemical company identified a waste water stream as a significant source of uncontrolled emissions. The waste water is generated by steam jet eductors from a vacuum column used in a chemical manufacturing process. The condensed steam from the jet eductors is contaminated with about 530 ppm of ethyl chloride and smaller quantities of butyl chloride, benzyl chloride and non-chlorinated organics, primarily toluene.

The waste water treatment project was on an extremely aggressive time line to meet corporate emission reduction deadlines. The project scope provided for the design, manufacture, and pre-assembly of a complete unitized, skid-mounted system in less than eight weeks to allow on-site installation, commissioning and start-up to be completed within four weeks.

Thermatrix designed, fabricated and supplied a 150 m^3 hr^{-1} electrically heated reactor as part of the work scope for this client. The reactor was integrated into an abatement system consisting of an air stripper, knock-out pot, flameless oxidizer, HCl scrubbing system and fully automated controls.

Approximately 11,000 kg hr^{-1} of waste water is admitted to the air stripping column that is designed to remove 99.9% of the volatiles and produce a moist air stream containing the organics. The cleaned water is recycled to the plant, while the 150 m^3 hr^{-1} stripper off-gas is conveyed through a knock-out pot and demister before entering the flameless oxidizer, where 99.99% destruction of the organics has been demonstrated to be achievable. The oxidation reaction produces CO_2, H_2O and HCl. Upon exiting the oxidizer, the gases are quenched and admitted to the scrubbing tower, where 99% of the HCl gas is removed. The scrubber water is discharged from the system to the plant waste water system and the organic-free and acid-free gases exit the scrubber to atmosphere.

To minimize the on-site work, the treatment system was designed and pre-assembled complete with all piping, instrumentation and electric power systems. On-site work only involved completing the few process piping tie-ins, terminating a single power feeder and multi-conductor control cable, and erecting the stripping and scrubbing towers which are too tall to be transported in place. Pile foundations, field piping and electrical runs and certain site improvements were completed while the system was being manufactured.

The system was installed, started up and commissioned without any significant delays. The air permit for the system was issued by state authorities in 30 days. Summary results from compliance testing are presented in Table 16.2.

TABLE 16.2
Treatment of VOCs from air stripping operations — summary of results

Permitted compounds	Permit limit (60 minute averages)	Measured actual emissions Run 1, 2, 3
Chloroethane	0.023 kg hr^{-1}	Run 1: 0.000450 kg hr^{-1} Run 2: 0.000407 kg hr^{-1} Run 3: 0.000439 kg hr^{-1}
Chlorobutane	0.0023 kg hr^{-1}	Run 1: 0.00118 kg hr^{-1} Run 2: 0.00103 kg hr^{-1} Run 3: 0.00106 kg hr^{-1}
Toluene	0.0023 kg hr^{-1}	Run 1: 0.00186 kg hr^{-1} Run 2: 0.00146 kg hr^{-1} Run 3: 0.00154 kg hr^{-1}
CO	0.027 kg hr^{-1} or 100 ppmv @ 7% O_2	Run 1: 0.00016 kg hr^{-1}, 8.7 ppmv Run 2: 0.00013 kg hr^{-1}, 7.1 ppmv Run 3: 0.00010 kg hr^{-1}, 3.1 ppmv
HCl	0.045 kg hr^{-1}	Run 1: 0.00037 kg hr^{-1} Run 2: 0.00037 kg hr^{-1} Run 3: 0.00049 kg hr^{-1}
THC (total hydrocarbon)	50 ppmv or 95% destruction and removal efficiency (DRE)	Run 1: 10.2 ppmv as methane Run 2: 4.8 ppmv as methane Run 3: 5.1 ppmv as methane Run 1: 0.00032 kg hr^{-1} as carbon Run 2: 0.00017 kg hr^{-1} as carbon Run 3: 0.00015 kg hr^{-1} as carbon Run 1: 99.97% DRE Run 2: 99.98% DRE Run 3: 99.99% DRE

ABATEMENT OF HALOGENATED VOCs FROM HERBICIDE PRODUCTION

A major chemical company installed and is operating a Thermatrix FTO system for the treatment of methylene chloride emissions from herbicide production. Prior to this installation, traditional flame-based technology was the corporate standard for this application. The herbicide manufacturing process consists of various unit operations that continuously or intermittently vent process gases containing CVOCs. The combined vent stream includes 125 kg hr^{-1} methylene chloride, 2.7 kg hr^{-1} CO and traces of methanol, formaldehyde and dichloromethyl ether. Venting results from equipment depressurization, controlled process venting, equipment purges, batch chemical transfers and normal breathing losses. Vents are collected and routed to the Thermatrix system for treatment.

The skid-mounted, fully automated abatement system consists of a Thermatrix reactor and an effluent gas quench which feeds directly to a pre-existing scrubber system. The system is designed for a total flow of 2500 m^3 hr^{-1} and a CVOC destruction and removal efficiency of 99.99%. The system is fed by two vent collection headers which are combined immediately prior to entering the main fume line. Both streams are water saturated, with one containing high concentrations of CVOCs inerted with nitrogen to reduce flammability. The second stream contains relatively low concentrations of CVOCs and is continuously purged with air.

During operation, combustion air is added to the combined vent stream in the main fume line to maintain a minimum oxygen concentration. The premixed fume is then introduced to the Thermatrix reactor, where organics are oxidized to CO_2 and H_2O vapour. An acid gas (HCl) is produced and quenched, then sent directly to a caustic scrubber for neutralization. Figure 16.4 (pages 194–195) presents the process flow diagram for the system.

Installation of the system was completed at the client's site in seven days in January 1995. During commissioning of the system, source testing was performed to ensure a system performance meeting local air board requirements. Inlet samples containing up to 300 ppmv of total hydrocarbons were taken from the main fume line. Outlet samples collected at the stack revealed undetectable hydrocarbons at a 1 ppmv detection limit, demonstrating compliance with local air board requirements.

In 1995, a large FTO based system was also designed, fabricated, installed and commissioned at a major herbicide production plant. The system is capable of treating 5000 m^3 hr^{-1} of chlorinated and non-chlorinated VOCs. The system is skid-mounted and major components include the flameless thermal

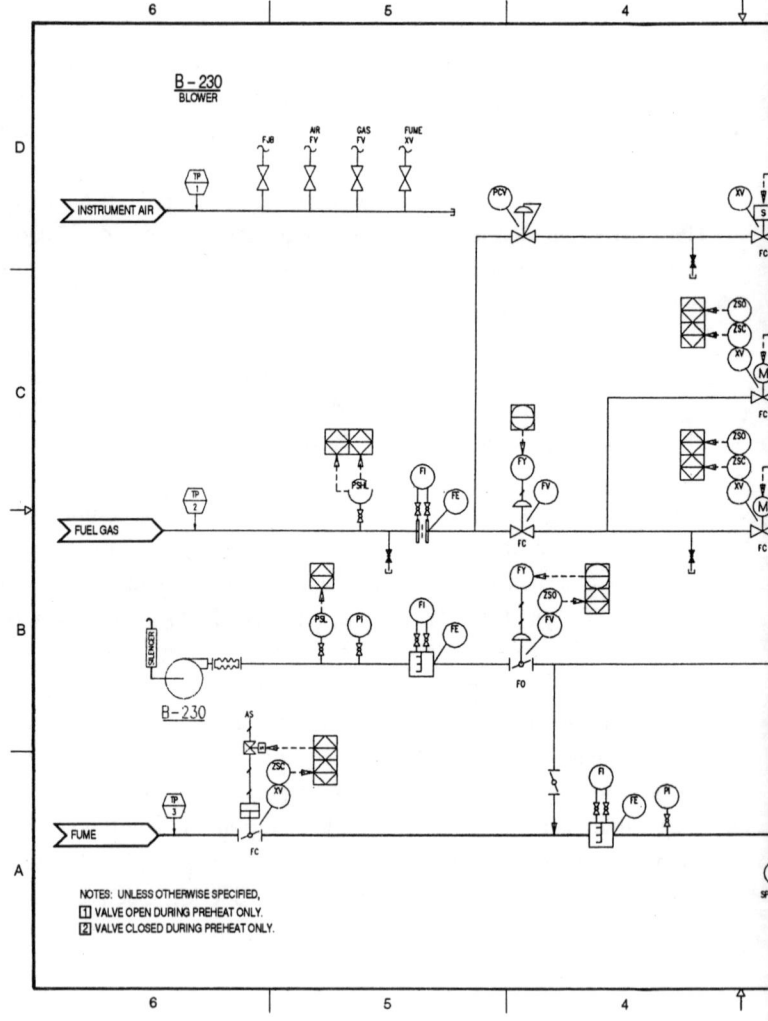

Figure 16.4   Typical Thermatrix FTO P&ID.

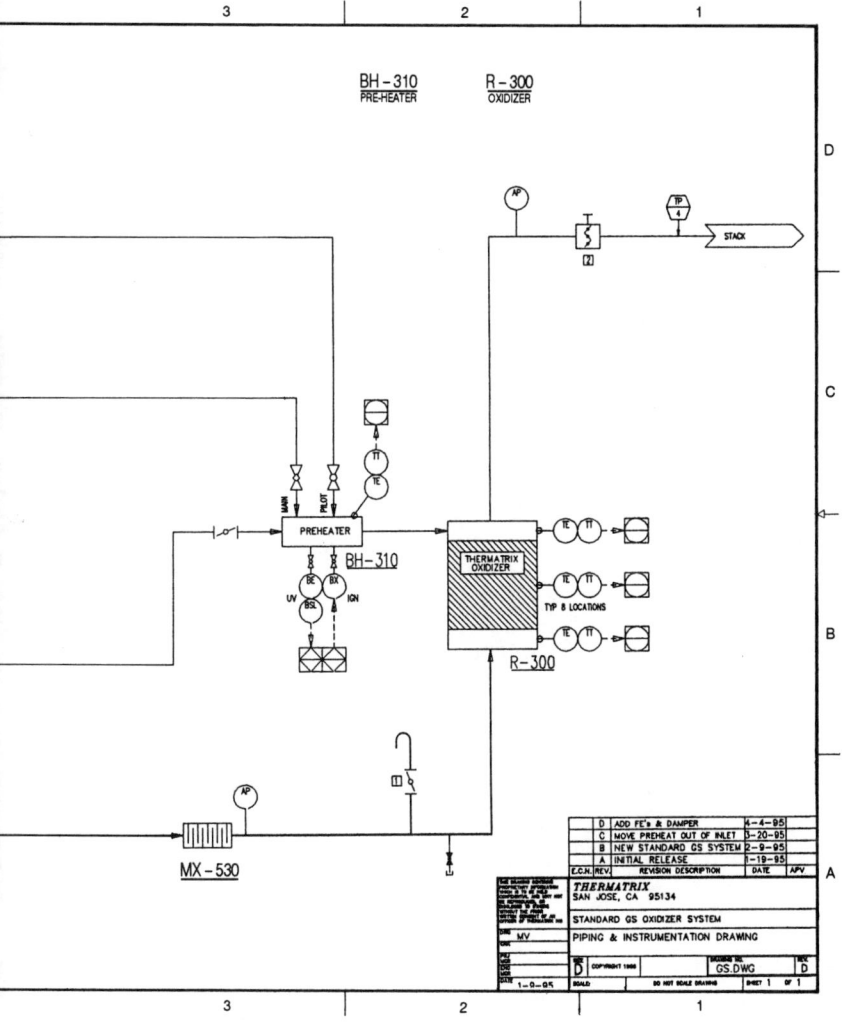

195

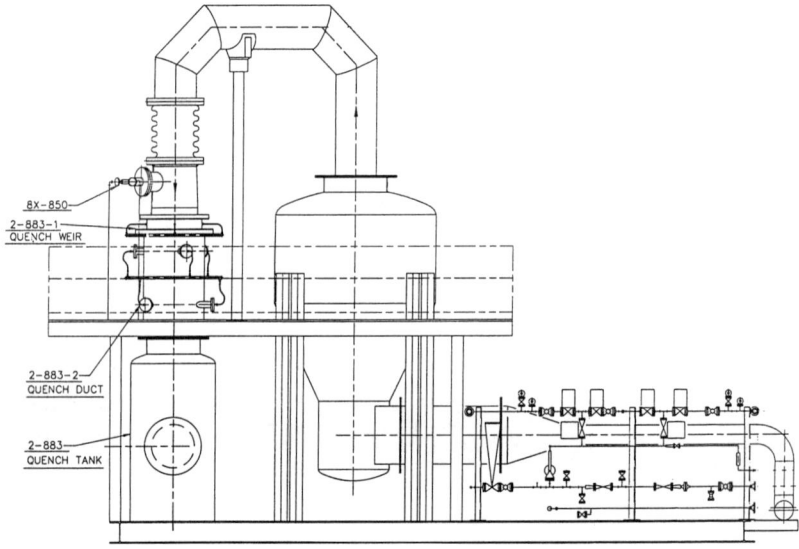

Figure 16.5   Typical layout of FTO and effluent gas quench unit.

oxidizer, quench/scrubber and stack. Figure 16.5 is a general arrangement drawing of the system. The system is currently on line and treating plant VOC streams.

TREATMENT OF CVOCs EXTRACTED FROM SOIL AT A CHEMICAL PLANT
A specialty chemical manufacturing company ordered and installed a gas straight-through 250 m^3 hr^{-1} FTO for use in the treatment of vapours extracted from CVOC contaminated soil at the plant. The system was commissioned and placed into operation in 1995. For compliance with local air regulations, a destruction and removal efficiency of > 95% is required. Test results show destruction and removal efficiencies in excess of 99.9% were obtained.

CONCLUSIONS
As demonstrated by the case histories presented above, flameless thermal oxidation has proven to be a successful technology in treating multiple process and remediation vapour streams which contain CVOCs. Very high destruction and

removal efficiencies have been demonstrated with flameless thermal oxidation of vapour streams containing CVOCs. The technology lends itself well to scale-up. Initial commercially available oxidizers had vapour stream processing rates of 2–8 m^3 hr^{-1}. In 1995, the successful commissioning and operation of FTOs with vapour stream treatment rates of 2500–5000 m^3 hr^{-1} was accomplished, and in 1996 Thermatrix technology successfully destroyed an off-gas containing chlorinated and fluorinated hydrocarbons.

REFERENCES IN CHAPTER 16

1.   Martin, R.J. *et al*, 1993, Selecting the most appropriate HAP emission control technology, *The Air Pollution Consultant*, 3 (2).

2.   Allen, M.W. *et al*, 1995, Flameless thermal oxidation of low concentration VOC remedial waste streams: designs for planned DOE demonstrations, *Waste Management 95 Conference, 26 February–2 March 1995, Tucson, Arizona, USA*.

3.   US Department of Energy, September 1995, *Flameless Thermal Oxidation, Innovative Technology Summary Report* (Office of Technology Development, US Department of Energy).

17. ODOUR REDUCTION USING ACTIVATED CARBON

Tim Kermeen

INTRODUCTION

This chapter discusses the removal of disturbing odours, mainly emitted from municipal waste water treatment plants. It looks at the alternative treatment processes, the use of activated carbon to eliminate odours, the equipment used and economic factors governing the selection of odour control systems.

Odour, as it is encountered in the Western world, can best be described by the word nuisance. This does not mean that what one person recognizes as a disturbing odour is perceived as such by other humans. To be precise, odour has three characteristics that identify it as an odour or a scent:

- detectability;
- intensity;
- the combination of quality and acceptability.

DETECTABILITY

In nature the nose and olfactory system are used in survival. Every animal knows by instinct that danger, food and a possible mate are announced by a special odour. Humans lead a more sheltered life, and for the most part have lost the use of this sense. Some instinct is left intact, of course, otherwise the perfume manufacturers would be out of business. When humans sense an odorant such as hydrogen sulphide, they know something is wrong, but do not necessarily take action to avoid or eliminate the source of the nuisance. One of the most difficult problems in promoting an awareness of how potentially dangerous these situations can be is that detectability is very personal and totally individual.

The best instrument for detecting odour is the human nose. It can detect a wider range of contaminants than any other instrument currently on the market. This is the reason why odours are commonly measured by an odour panel. The dilution at which 50% of the panel can smell the odour is called the number of odour units. The higher the odour number, the greater the odour nuisance. Another reason for the use of a panel is that, most of the time, odour consists of many components; detecting only one does not necessarily describe the extent of the nuisance.

TABLE 17.1
Sources of odour

Compound	Detection threshold level, ppb(v)
Skatole	0.06
Hydrogen sulphide	0.47
Indole	1.00
Methyl mercaptan	1.10
Methyl amine	21.00

Table 17.1 lists some of the more common odorants encountered (in a negative sense, that is) and their detection limits.

Another value related to this point is the threshold limit value (TLV). This is defined as the time-weight average concentration for a normal 8-hour working day or 40-hour working week to which workers, presumed to be healthy, are exposed day after day without adverse effects. Lists of these TLVs should be reviewed regularly, taking account of new detection techniques and field data.

INTENSITY

Intensity is linked with another parameter called adaption. Humans with a normal sense of smell can be subject to a phenomenon called olfactory fatigue or self-adaption to the odorant. This phenomenon appears when the concentration of the odorant is fairly constant. The perceived intensity of the smell fades and this is described as human self-adaption. The real intensity is not detected, nor is there warning of any potential danger. The human sense of smell is, like many other senses, related to the power law. In the case of smell this law is Steven's psychophysical power law:

$$I = K \times C^n$$

where:
I = perceived intensity;
K = unit related constant;
C = concentration of the odorant;
n = logarithmic slope.

This formula explains why the perceptibility of certain products (with a low 'n') is less dependent on their concentration than those of other products (with a high 'n'). In other words diluting helps to get rid of the latter products but does not help to remove the stench of the first type.

QUALITY AND ACCEPTABILITY

Quality and acceptability are mostly dependent on the culture and the attitude of the person detecting the odour. In other words people classify a smell as being a fragrance, an aroma, a neutral or a malodorous stench, depending on their background. Testing a certain smell for quality before a panel causes so many variances that regional parameters must always be taken into consideration.

The odours produced by municipal or industrial treatment works are generally from three sources:

- decaying organic matter;
- volatile organic compounds (VOCs);
- bacteria (principally anaerobic H_2S-forming bacteria).

Table 17.2 gives an overview of those products that are most prominent in treatment works and so are a prime target for odour control. They are easily divided into two groups — sulphur and ammonia.

TABLE 17.2
Sources of odour often found in waste water treatment works

Compound	Typical formula	Description
Amines	$CH_3 (CH_2)_n NH_2$	Fishy
Ammonia	NH_3	Ammoniacal
Diamines	$NH_2 (CH_2)_n NH_2$	Decayed fish
Hydrogen sulphide	H_2S	Rotten egg
Mercaptans	$CH_3 (CH_2)_n SH$	Skunk secretion
Organic sulphides	$(CH_3)_2S$ CH_3SSCH_3	Rotten cabbage
Skatole and indole	$C_8H_5 NH CH_3$ $C_8H_6 NH$	Faecal

Although these products may appear at the same plant their nuisance value is not the same. The odour threshold can be spread over several different orders of magnitude, so the prime point to be aware of is that removing a highly detectable odour makes less detectable odours more prominent. Therefore the production of odour-free air is not defined by the removal of the most disturbing component (see the odour threshold limits in Table 17.1).

HYDROGEN SULPHIDE

Hydrogen sulphide (H_2S) — a colourless but certainly not odourless substance — is probably the best known pollutant and source of odour nuisance. It is produced in almost all sewage works.

H_2S is easily detectable at levels below 1 ppb, but at these concentrations it is perceived as a sweet odour. At ppm level it becomes the familiar rotten egg stench and once above 100 ppm all sense of odour is lost and exposure is potentially fatal.

The serious odour problems associated with the collection, handling and treatment of primarily domestic waste water are mainly the result of the reduction of sulphate to H_2S under anaerobic conditions. Sulphate is one of the most universal anions occurring in all natural waters, and in rainfall (especially over urban areas). The addition of human and animal excreta to surface water from our domestic waste water means that the anaerobic and facultative bacteria have an excellent food source for the production of H_2S.

In general, reducing the concentration of H_2S also reduces the concentration of related sulphur-containing organics. If carbon is used as the odour control agent, the more volatile organic odorants can be removed as well.

Figure 17.1 shows the rise in perceived odour within a typical waste water treatment plant, from initial receipt of the waste water into the process sump and throughout the various treatments (all H_2S measurements are on a static basis).

TECHNOLOGIES USED IN ODOUR CONTROL

CARBON ADSORPTION

This technology is dealt with in the next section.

WET SCRUBBING

Wet scrubbing is an excellent technology for handling high loads of H_2S. It does not, however, have the ability to respond easily or quickly to peak loads. Most

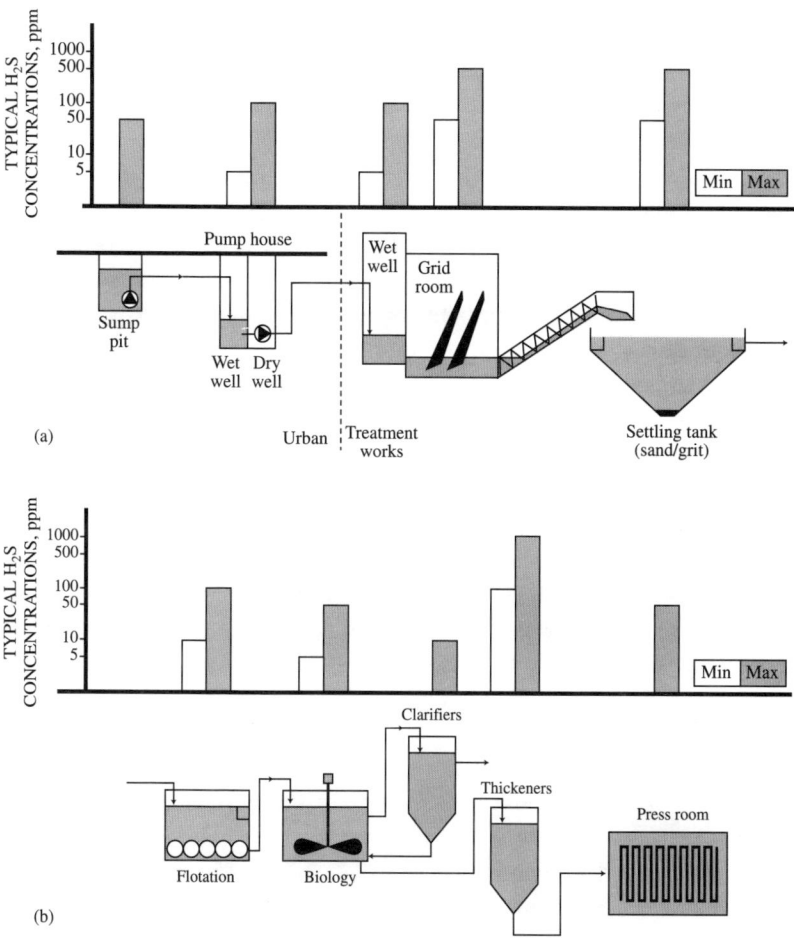

Figure 17.1   Municipal odour control — potential odour areas.
(a) Initial treatment for solids removal; (b) Activated sludge treatment.

scrubbers are designed around a set point for inlet odour and chemical dosage. To achieve high removal efficiencies requires multistage scrubbers with two or three set points and a consequent increase in capital cost. An odour-free treated stream cannot be guaranteed.

All this makes wet scrubbing an ideal candidate for primary air treatment but a polishing stage must follow.

IMPREGNATED ACTIVATED ALUMINA

Activated aluminas are specific for one odour — H_2S. They have no adsorption capacity for organic odours and in view of the range of odorants normally found together, they are highly unlikely to produce odour-free air.

BIOFILTERS/PEAT BEDS

Biofilters are the 'eco-solution'. There are many types and they are economical except in terms of space. Their chief disadvantage is a sensitivity to variations in the feed stream in terms of both concentration and the type of inlet compounds. Biofilters are best suited to constant feed conditions that can maintain a stable biomass; they struggle to handle peak loads. More importantly the treated air is never odour-free, but they do provide acceptable results in the right location and with the help of dilution.

MASKING AGENTS

Obviously masking agents do not attack the root problem of odour — depending upon individual perception they can even make the odour worse. They tend to be expensive and some evidence suggests that water spraying can be equally effective.

CARBON ADSORPTION

In order to achieve real odour-free air, organics and volatile odorants such as H_2S must be removed. All activated carbons do this to some extent, but not all carbons are the same. There are three basic types:

- non-impregnated activated carbon;
- alkali-impregnated carbon;
- Centaur carbon.

  Normally all activated carbons adsorb odorants. Plain activated carbon can function both as an adsorbent and a catalyst in the control of odours. Organic products are physically adsorbed and sulphur-containing products such as H_2S and mercaptans are oxidized on the carbon surface. Major factors affecting adsorption can be summarized as:

- activated carbon type;
- nature of contaminant;
- concentration of contaminant;
- pressure;
- temperature;
- relative humidity.

When selecting a carbon for odour control, it is a good idea to take all of these factors into account. The best system designs rely on pilot data and case histories.

All activated carbons, in the presence of moisture, act as oxidation catalysts and catalyse the oxidation of H_2S to elemental sulphur. The sulphur is then adsorbed onto the carbon surface. Generically the mercaptans are oxidized to less odorous and more retainable disulphides. The capacity of an ordinary carbon for these reactions is very low; the retention of these odorants can, however, be enhanced by the addition of impregnants to the activated carbon.

IMPREGNATED CARBON

The impregnated carbon type IVP is a coal-based high activity vapour phase carbon impregnated with sodium hydroxide. This combination acts not only as an adsorbent for organic pollutants, but also as an efficient desulphurization agent. The impregnation of sodium hydroxide means that IVP can load up to 25% by weight of H_2S instead of the usual 1–2% experienced with non-impregnated carbon.

Figure 17.2 shows the ways in which the hydrogen sulphide is chemically converted into better retained and less odorous products. These products are mainly elemental sulphur and sodium sulphide.

Another parameter which has to be taken into account is that this product can eliminate odorants, via the above processes, down to almost non-detectable levels. Due to the synergistic effect of the impregnant and the activated carbon, this can be achieved using 'normal' contact times. This is very important because it influences dramatically the size of the equipment, and hence the capital investment in the odour abatement system.

$$H_2S \text{ (gas)} \xrightarrow[\text{Pore structure}]{\text{GAC}} H_2S \text{ (liquid in pores)}$$

$$\xrightarrow[\text{NaOH/Na}_2\text{CO}_3]{\text{Acid/base reaction}} Na_2S + H_2O + H_2CO_3$$

$$\xrightarrow[O_2]{H_2O} S \text{ (elemental)} \quad \text{Typical loading} <25\%$$

Figure 17.2 Hydrogen sulphide removal.

Caustic impregnated carbon can be regenerated *in situ*, for multiple cycles. Regeneration requires 50% caustic soda solution and takes approximately three days.

CENTAUR CARBON

Centaur is a new type of activated carbon. Put simply, it has the performance of an impregnated carbon without the need for controlled disposal, or the potential hazard of a lowered ignition temperature. Centaur has the same pore structure as normal activated carbon, which means that it can adsorb organic products as easily as a plain carbon. All impregnated carbons lose adsorption capacity because of the chemical impregnation that obstructs their available adsorption area.

The primary advantage of Centaur is that, while a normal carbon has a limited range of catalytic activity, Centaur has the capacity to enhance a wide range of reactions to a significantly greater extent.

The catalytic activity of an activated carbon is linked to its electronic structure. The electronic structure of the graphite platelets found in all activated carbons is generally capable of handling a small number of reactions. With Centaur the number of catalytic sites is increased by a factor between ten and 100; consequently uneconomic or minor side reactions possible on carbon now become significant.

The mechanism of reaction is initially similar to that of all activated carbons and the contaminants are adsorbed within the structure of the carbon. These contaminants are chemically-changed due to the presence of a catalytic initiator. Some of these products can then be removed during a regeneration step. Regeneration of Centaur carbon is achieved by a simple water wash which removes oxidized H_2S as sulphuric acid.

The capacities of the three types of carbon reviewed here can be summarized as follows. A plain activated carbon possessing some catalytic activity will achieve a maximum loading of approximately 1–2% for hydrogen sulphide.

With caustic impregnated carbons an acid-base reaction follows physical adsorption and the formation of elemental sulphur. Under normal circumstances this carbon can be regenerated by caustic soda solution. The initial loading of hydrogen sulphide can be as high as 25% by weight; this is dependent, however, upon the mass of organics present. The catalytic reaction occurs but contributes only 1–2% of the total loading for H_2S.

Centaur takes these reactions a stage further; elemental sulphur is converted to sulphuric acid, which can be removed by a water wash. With Centaur

TABLE 17.3
Comparison of activated carbons

Specification	BPL[a]	IVP[b]	Centaur®	Coconut
Iodine number	1000	–	800	1250
Apparent density, g ml^{-1}	0.52	0.60	0.56 (min)	0.48–0.52
Particle size	4×6	4×6	4×6	4×6
MPD[c]	3.6	3.6	3.6	3.6
TM–41R (g H$_2$S ml^{-1} carbon)	0.01–0.03	0.14	0.09–0.10	0.02

[a] BPL — coal-based gas phase activated carbon
[b] IVP — impregnated coal-based gas phase activated carbon
[c] MPD — mean particle diameter

an initial loading of around 20 wt% is possible, depending on the mass of organics present.

When comparing a catalytic carbon (Centaur) with a chemi-adsorption carbon such as an IVP, there are a number of additional factors besides the issue of capacity. Centaur has no impregnation and hence more free volume for adsorption. Its ignition temperature is higher than any of the impregnated carbons, regeneration is simpler and cheaper (requiring only water) and is less time-consuming. A typical water regeneration cycle lasts a day rather than the three days required for caustic rinse.

Lastly, at the end of its working life catalytic carbon can be thermally regenerated, whereas impregnated carbons require controlled disposal.

Table 17.3 gives the specifications of the carbons discussed.

ODOUR CONTROL EQUIPMENT SIZING AND DESIGN

The range of flows that can be treated by single odour control units varies from a few hundred cubic metres per hour for the deodorization of small pumping stations, up to thirty thousand cubic metres per hour for the deodorization of filter press rooms. The size of equipment depends primarily on two interlinked parameters: the superficial or empty bed contact time (EBCT) and the linear velocity. These two factors define the type of odour control unit (OCU) required. Other key parameters which influence equipment choice are carbon usage and the availability of plant space.

The EBCT defines the size of the carbon bed and must fulfil two criteria:
- to achieve effective odour removal, carbon beds require sufficient contact time with the process air;
- the bed should have an acceptable design life between regenerations.

The equipment must have:
- even air distribution;
- air monitoring points;
- regeneration pipework;
- the required air changes per hour for effective site odour control.

The last point is crucial in the efficient design of an odour control system. The best way of preventing odour escape is by creating an underpressure in the odorous area. In reality this is very difficult to achieve but to establish the number of air changes per hour (AC/h) the following formula can be used:

$$AC/h = 2.5 \times F_1 \times F_2 \times F_3 \times F_4$$

where:

$$Flow = volume \times AC/h$$

F_1 = climate factor;
F_2 = pollution level;
F_3 = the treatment objective;
F_4 = the leakage factor or sealing of the building or vessel under control.

Naturally the higher the level of pollution, the hotter the climate, the lower the treatment objective and the poorer the vessel sealing, the greater the number of air changes required.

Closely monitored tests in Dubai had the following results. A Ventsorb 70 containing 75 kg of IVP carbon was 202 days on line producing odour free air at a flow of 175 m^3 h^{-1}. During this period the H$_2$S inlet concentration varied between 2 and 450 ppm (ratio 1:225). Relative humidity fluctuated between 12 and 100% and the temperature from 19 to 45°C.

During this odour-free period the total loading of H$_2$S on the IVP activated carbon added up to 27% on a weight basis. A number of other odorous products were also adsorbed, but not analysed. Hydrogen sulphide, being the smallest molecule and the best detectable product, was the first to break through.

SUMMARY

Peat bed biofilter systems, wet scrubbers and regenerable carbon systems have been examined to compare economic performance.

Each technology has its distinct advantages depending on circumstances, but an analysis of capital and operating costs over a ten-year period raises a number of points. Peat beds are linear from an economic perspective — doubling flow effectively doubles size and cost. Scrubber technology, on the other hand, faces disproportionately high capital cost for smaller units. Carbon represents a balance between these two technologies, offering either a polishing stage to produce odour-free air or a stand-alone odour control system in its own right.

REFERENCES IN CHAPTER 17

1.   Water Pollution Control Federation, Odour control for wastewater facilities, *Manual of Practise Nr 22*, 77 (Water Pollution Control Federation, Washington DC, USA).

2.   Committee on Medical and Biological Effects of Environmental Pollutants (subcommittees on H2S), *Hydrogen Sulphide*, 79 (Division of Medical Sciences, National Research Council, Washington DC, USA).